BRAND-NAME HANDBOOK
of Protein, Calories,
and Carbohydrates

Also by the Author:

All Natural Pain Relievers

Health Secrets from the Orient

Health Tonics, Elixirs and Potions for the Look and Feel of Youth

Magic Enzymes: Key to Youth and Health

The Miracle of Organic Vitamins for Better Health

Miracle Protein: Secret of Natural Cell-Tissue Rejuvenation

Natural and Folk Remedies

Natural Hormones: The Secret of Youthful Health

The Natural Laws of Healthful Living: The Bio-Nature Health Rhythm Program

The New Enzyme Catalyst Diet: Amazing Way to Quick, Permanent Weight Loss

BRAND-NAME HANDBOOK
of Protein, Calories, and Carbohydrates

*Carlson
Wade*

Parker Publishing Company, Inc.
West Nyack, New York

Library of Congress Cataloging in Publication Data

Wade, Carlson.
  Brand-name handbook of protein, calories, and
carbohydrates.

    1. Food--Protein content--Tables. 2. Food--
Caloric content--Tables. 3. Food--Carbohydrate
content--Tables. I. Title.
TX551.W16   641.1   76-47457
ISBN 0-13-081307-9

DEDICATION

To the New Health You Will Enjoy

FOREWORD
by a Doctor of Medicine

Here at last is an all-inclusive, health-building, encyclopedic reference that takes away all the mystery about the foods you see on the shelves of your local market.

Carlson Wade, an internationally recognized nutrition authority, has prepared a comprehensive, valuable book that gives you all the facts, figures, and advice as well as step-by-step programs you need to know in order to select deliciously healthy foods from the thousands of varieties currently available.

This nutrition handbook makes it easy for you to count three essential nutrients—*protein, calories, carbohydrates*—at a single glance! (No other book to my knowledge offers this three-way nutrition counter.)

To help you plan tasty and nutritious meals, Carlson Wade has prepared a brilliant new P/C/C Computer. This tells you *instantly* if the listed food is high or low in either of these three essential nutrients.

This means if you are on a high- or low-protein, high- or low-calorie, high- or low-carbohydrate program, just look under the P/C/C Computer column to decide if the selected food fits into your plans. *This takes all the guesswork out of diet planning.* This easy-to-read computer makes Carlson Wade's book so valuable for everyone who wants to enjoy youthful health of body and mind.

This book can help you succeed with any diet you are following. It will answer such typical questions as:

*Which foods are high in protein but low in calories?

*How can I enjoy low-calorie snacks that are also high in carbohydrates for speedy and all-natural energy?

*How can I easily plan a reducing diet with brand-name foods that do not always list nutritional data on their labels?

*What is the difference between "high" and "low" carbohydrate counts in fresh and processed foods?

7

*How can I buy, store, and cook fresh, frozen, and canned foods?

*How do I read labels?

These are just a few of the questions answered by Carlson Wade in his amazingly simple yet highly effective nutrition counter book. *It offers you more benefits than any other book because it gives you BOTH brand name and fresh food counts for all three essential nutrients.*

And, as explained above, a *bonus* feature is the priceless P/C/C computer that is unavailable anywhere else. It helps make nutrition planning an "instant" convenience.

Essential to calorie-conscious eaters is the need for a balance of protein and carbohydrates. It is true that you can have a youthfully slim shape on a low-calorie program. But you may be cheating your health on a low-protein or low-carbohydrate program. Just as it is important to count calories, it is also important to count other nutrients so that you are *nourished* while you are reducing.

Happily, Carlson Wade's book is the answer to this vital need. It tells you how to select low-calorie but adequate protein and carbohydrate foods for a delicious, healthy, and slimming result. No other book on today's market gives you this compendium of facts in such easy-to-read form. It makes dieting a pure joy!

Here, too, are included Official U.S. Government Recommendations for the need of protein, calories, and carbohydrates. You are told by the government not only what these nutrients are but what they can do in your body to build and rebuild your vital organs and processes and help give you an extended lease on a long and healthy life.

Spotlighted are the government's latest findings on the importance of these three nutrients and their official guidelines for use in daily eating.

You are also given the official U.S. Government's "Nine Steps to Controlled Calories" for weight reduction. Then you are given the Government's suggestions for using protein and carbohydrates for easy weight loss, healthy weight maintenance, and—even more important—the "easy way to plan a delicious daily diet" with four listed groups of foods. Clear. Simple. Health-rewarding.

Carlson Wade has sifted through the maze of currently available nutritional data. He has prepared a remarkably helpful book. It is easy to read, easy to follow, and highly rewarding in terms of youth-restoring health.

You are given the nutritional value of nearly 3000 brand-name and fresh foods. With this information, you can succeed with any diet you are following now or plan to follow.

Foreword by a Doctor of Medicine

Carlson Wade is to be highly recommended for having written the most valuable nutrition book of our modern times. Every home should have it!

William S. Keezer, M.D.

What This Book Can Do for You

This book was written for the purpose of giving you a comprehensive guide to the three most important nutrients found in everyday foods—calories, protein, carbohydrates.

It gives you the specific counts for these basic nutrients as they occur in nearly 3000 foods! This eliminates your guesswork in nutrition planning for yourself and your family.

Walk through any food market. You see fresh foods, processed foods, canned foods, frozen foods, dehydrated foods, junk foods. Which to select? You have entered a maze of confusion. You want calorie, protein, carbohydrate information on *all* of these foods so you can plan for healthier eating programs.

That is why this book was written. It is an easy-to-follow road map that takes you out of the food market maze and gives you "instant," at-a-glance nutritional counts for thousands of popular (and less popular) fresh and brand-name foods.

It shows you how to plan a high- or low-calorie, protein, carbohydrate program to rejuvenate your health from head to toe.

No longer are you compelled to select foods without sufficient knowledge of their nutritional counts. This book gives you many charts and tables that tell you exactly what is in the foods you eat. Now you can choose a wide variety of favorite fresh or brand-name foods and enjoy delicious taste together with good health because of these all-inclusive charts.

This book is more than a handbook of nutritional tables on the three most important health-building nutrients. It is a personal guide that shows you *how to plan your meals with these nutrients for improved health for yourself and your family.* You are also given the official U.S. Government recommendations for each of these three nutrients ... and told how you can use them for the building and rebuilding of youthful health.

More benefits include features on how to read food labels, how to understand weights, measures, and metric figures. You are told the meaning of those strange codes, facts, and figures that appear on labels. The mystery, the guesswork, the confusion of label reading is taken away with the clearly explained illustrative charts in this book.

A special bonus is the Protein/Calorie/Carbohydrate Computer which is *not* found in any other book. This easy-to-read computer (copyrighted exclusively for this book) tells you *instantly* whether the food is high or low in these nutrients.

Are you on a special diet which calls for regulating any or all of these nutrients? Just select your food, look it up in the tables, note the Computer column, and quickly discover whether it fits into your dietary plan.

The P/C/C Computer is a copyrighted, quick-simple rating system that immediately tells you which foods are high or low in the three essential nutrients. This makes it a joy in food planning.

Puzzled by the problem of planning nutritious meals with fresh, frozen, and canned foods? Then you will welcome the features in this book that tell you how to buy these foods, how to store them, how to cook them, how to get top nutritional benefits for your food dollar.

Whether you use everyday fresh foods or a variety of frozen or canned foods, you can easily plan good health with the valuable guidelines given to you in this book.

No need to make any guesses about nutrients you are (or are not) getting. Whatever diet you now follow, whatever diet you plan to follow, this book gives you the nutritional value in nearly 3000 generic and brand-name foods ... all at a glance!

There is *no* substitute for having these counts right in front of you. There is *no* substitute for knowing *how to plan* your protein, calorie, and carbohydrate program. This book shows you how to do this and much, much more. It is your doorway to youthful health through nutritional knowledge. Turn the doorknob and enter.

Carlson Wade

Contents

Brand Name Listing of Protein, Calories, and Carbohydrates (Including Basic Foods) (Cont'd.):

1 / How to Rejuvenate Your Body and Mind with P/C/C Planning

You can tap the hidden wellsprings of cellular-cerebral rejuvenation within your own body and mind with a unique, balanced daily intake of three basic nutrients. These three nutrients hold the key to the forces of self-regeneration that lie dormant in the physio-metabolic networks of your body. These three nutrients are able to unlock the inner reserves of self-perpetuation of the billions of body cells and tissues from your head to your toes. These three life-extending nutrients are *protein, calories,* and *carbohydrates.* When fed to your body in a proper balance, they are able to ignite the spark of self-replication, or repair, of almost every single one of your billions of cells. This process promotes a feeling of well-being and the appearance of youthfulness in your body and mind. *The key is to feed yourself a balance of protein, calories, and carbohydrates by easy-to-follow meal planning for good nourishment and an ample supply of "sparks" to ignite the inner flame of eternal youthfulness and good health.*

HOW THE P/C/C PROGRAM CAN BE THE KEY
TO PERPETUAL HEALTH

P/C/C: Key to Youthful Health. These three nutrients form more than the basis of life. Protein, calories, and carbohydrates will nourish your cells and provide the mechanisms whereby the body is alerted to carry out its important metabolic processes. But protein, calories, and carbohydrates need to be made available in *a proper balance* so that they can promote the release of energy within your body and enter into the process of tissue-building. If you are low in protein but high in calories, your tissues will break down while you gain excess fat. If you are high in protein but very low in carbohydrates, you subject your kidneys to an excess of acid because of this imbalance. If you are average in protein and carbohydrate intake, but low in calories, you will enjoy partial youthfulness but a decline in energy and vitality. The key to youthful health lies in maintaining a proper balance of these three nutrients. This book will show you how to do so with everyday foods, whether fresh, frozen, or canned. Let us take a closer look at protein, and also calories and carbohydrates, to see how their proper balance holds your youthful health in the balance, so to speak.

PROTEIN: STAFF OF LIFE

The foundation for youthful health is protein. Here is the official U.S. Government report[1] on the vital importance of protein:

"You are looking at a superb package of proteins when you see yourself in a mirror. All that shows—muscles, skin, hair, nails, eyes—are protein tissues. Teeth contain a little protein.

Unseen Body Protein. "Most of what you do *not* see is protein, too—blood and lymph, heart and lungs, tendons and ligaments, brain and nerves, and all the rest of you. Genes, those mysterious controllers of heredity, are a particular kind of protein. Hormones, the chemical regulators of body processes, and enzymes, the spark-plugs of chemical reactions, also are proteins. *Life requires proteins.*

---

[1]*Food, Yearbook of Agriculture;* U.S. Department of Agriculture, Washington, D.C., 1959. Pages 57-64.

VCRs that can read; TV sets for the wrist...

Here's a look at some of the coming breakthroughs in video technology

By David Lachenbruch

Computer TV. Within the next few years, you'll probably own a television computer instead of a TV set. Virtually all set makers are employing increasing numbers of microcomputers in their products. The German manufacturer Standard Elektrik Lorenz recently showed a set in which all of the picture and sound circuits were computerized, and demonstrated some features that could be included. By pushing a button on the wireless remote control, the viewer can "freeze" any picture for detailed inspection, going back to the regular TV picture by pressing the button a second time. In Europe, where TV-set flicker is a problem, the set provides a rock-steady picture by increasing the number of times the picture changes from 50 per second to 75. In the U.S., the set could automatically eradicate ghosts and interference and eliminate the picture's line structure. In one version of the set, the viewer could watch several channels at a time, placing the pictures side by side on the screen. Standard Elektrik plans to phase in computerized sets in the next three years, and set makers in the U.S. and Japan are already working on similar designs.

High Definition. A new television service with wide-screen pictures of CinemaScope proportions, movie-theater sharpness, enhanced color and stereophonic sound is being proposed by television engineers, including those of ABC and CBS. CBS, in fact, has held several demonstrations of the improvements possible in the TV picture, showing special wide-screen tapes of part of an NFL football game, the 1982 Tournament of Roses Parade and some segments of *The Fall Guy*, produced by Glen Larson. High-definition TV, or HDTV, would increase the horizontal scan-line standard from 525 lines to more than 1000 and thereby produce pictures greatly improved in clarity. Proponents of HDTV argue that the coming allocation of a new band of satellite frequencies will offer the last chance in this century for improved TV, and they're trying to persuade the Government to keep this band open for deluxe HDTV rather than earmark it for more channels of today's type of TV. Under various proposals, HDTV could be beamed terrestrially from station transmitters, directly from satellites to homes or a combination of both—possibly within the next decade.

VCR That Can Read. Presetting a "programmable" videocassette recorder to record several different →

continued

shows on different channels is simple—theoretically—but it's time-consuming, and even experienced video engineers have been known to have their machines tape shows at 4 A.M. when they meant to punch in 4 P.M. Seeking to make the process simpler and quicker, a German manufacturer and a publisher have jointly developed an electronic gadget that can read TV program listings and remember to turn a home VCR on to tape them. Each listing in the publisher's program guide is to be followed by a "bar code" similar to those used on products in the supermarket. A pen-sized scanner attached to the programmer (which plugs into the VCR) is passed across the code, and the recorder then is automatically set to record that program when it comes on and to turn off when it's over. The gadget can preset a programmable VCR to record up to four programs on different channels.

Video Snapshots. Several Japanese manufacturers are working on still cameras that use magnetic materials instead of film. Sony's Mavica resembles a 35mm SLR camera and stores up to 50 color snapshots on a magnetic disc, about the size of a Ritz cracker, inside the camera. The pictures can be played back through a special viewer attached to the TV set. A 50-picture disc can be mailed in an envelope; or if you want to get the picture to Grandma more quickly, it can be sent by telephone, using an adapter. Sony hopes to introduce the system in the United States next year at about $1000. The company will also market a device to make prints from the disc, but the electronic prints may not have the detail or clarity of a film print.

The Big Picture. You can look for a breakthrough in more modestly priced giant-screen television later this year with the introduction by several manufacturers of reasonably compact consoles with screens measuring 40 to 50 inches diagonally. Some of them require no more floor space than 25-inch sets—which they resemble—but they have three to four times

the picture area. Instead of a direct-view picture tube, they'll use a new-type projection system that throws a bright, sharp picture from the inside of the set onto the face of a translucent screen. This is the same principle used in bulkier projection sets, but the cabinets have been shrunk to console size by the use of a new miniature optical system. The price should be around $2000—not exactly cheap, but about $1000 less than today's heftier television projectors.

Paging Dick Tracy. An "information system for the wrist" is being developed by Seiko, the big Japanese watchmaker. Although the project is shrouded in secrecy, it's known that the company has developed a tiny TV set using a liquid-crystal display of the type employed in digital watches. There's speculation that the "information system" may combine TV, calculator and electronic "memo pad," perhaps with a tiny video game thrown in for good measure. And—oh yes—it also tells the time.

Watchman. Beware of walking viewers when Sony introduces a tiny flat-screen portable TV set later this year. Possibly inspired by the success of its Walkman portable stereo cassette player that, with its imitators, has aurally isolated thousands of otherwise respectable citizens, Sony's new portable is smaller than a paperback book, weighs just over a pound and can be viewed on the walk, so to speak. The secret of its svelte figure (it's only 1¼ inches thick) is a new 2-inch picture tube whose stem is parallel to the viewing surface instead of being at right angles as in conventional tubes. The little black-and-white set operates from four AA batteries that provide 2½ hours of continuous viewing. Already on sale in Japan at about $240, the set probably will cost slightly more here. Another set of similar design, but with a giant 3-inch screen, is promised here this year from a British company, Sinclair Research Limited. Sinclair says its tiny portables will sell for about $100. (END)

Two new ways to play it...

New Kool Lights
There's only one
low 'tar' with
a sensation
this refreshing.
Kool Lights.
The taste doesn't
miss a beat.

New Kool Ultra
One ultra has
taste that outplays
them all.
New Kool Ultra.
Even at 2 mg., you
get the refreshing
sensation of Kool.

A STAR AT 6...
MOTHER AT 15...
DIVORCED AT 16

Lisa Loring's real-life story is as sudsy as her role in 'As the World Turns'

By Robert Bridges

In oversized rollers and knockabout corduroy pants, Lisa Loring nervously prowls the halls of CBS Studio 52 in New York, complaining about her dentist. "If I'd known how much he planned to do, I'd have waited for my day off," she moans, peeling back her lower lip to reveal tiny black stitches. She must suffer, if not silently, the results of her gum surgery, because within the hour Loring will be starting work—ruining romances and stirring up inordinate amounts of trouble as Cricket Montgomery, one of *As the World Turns'* sleek young vixens.

Loring is capable of creating theatrical

28

Source of Life-Giving Proteins. "Proteins have to be made by living cells. Proteins do not exist in air, like nitrogen or oxygen. They do not come directly from the sun, like energy. We must get our proteins from plants and other animals. Once eaten, these proteins are digested into smaller units and rearranged to form the many special and distinct protein components we need.

Protein in Your Body. "Next to water, protein is the most plentiful substance in your body. If all the water were squeezed out of you, about half your dry weight would be protein. About a third of the protein is in your muscle. About a fifth is in your bone and cartilage. About a tenth is in your skin. The rest is in your other tissues and body fluids.

Protein in Your Bloodstream. "There are several dozen proteins in your blood alone. One of the busiest is hemoglobin, which constantly transports oxygen from your lungs to your tissues and brings carbon dioxide back from your tissues to your lungs. Ninety-five percent of the hemoglobin molecule is protein. The other 5 percent is the portion that carries the iron. Other proteins in your blood are defenders, for they give you the means of developing resistance and sometimes immunity to disease. Gamma globulin (a protein) can also form antibodies, substances that can neutralize bacteria and viruses and other micro-organisms. Different antibodies are specific for different diseases. Gamma globulin also helps the scavenger cells—phagocytes—engulf disease microbes.

Maintains Water Balance. "Proteins help in the exchange of nutrients between cells and the intercellular fluids and between tissues and blood and lymph. When you have too little protein, the fluid balance of your body is upset, so that your tissues hold abnormal amounts of liquid and become swollen.

Daily Protein Needed for Cellular Growth. "The proteins in your body tissues are not there as fixed, unchanging substances deposited for a lifetime of use. They are in a constant state of exchange. Some molecules or parts of molecules always are breaking down and others are being built as replacements. This exchange is a basic characteristic of living things; in your body it is referred to as the dynamic state of body constituents—the opposite of a static or fixed state. *This constant turnover explains why your diet must supply adequate protein daily even when you no longer need it for growth.* The turnover of protein is faster within the cells of a tissue (intracellular) than in the substance between the cells (intercellular).

How Protein Provides Youthful Energy. "One gram of protein will yield about 4 calories when it is combined with oxygen in your body. One ounce will give you 115 calories. That is about the same amount as starches and sugars give. *Your body puts its need for energy above every other need. It will ignore the special functions of protein if it needs energy and no other source is available.* This applies to protein coming into your body in food and to protein being withdrawn from the tissues.

Process of Energy-Creating Protein. "Either kind gets whisked through your liver to rid it of its nitrogen and then is oxidized for energy without having a chance to do any of the jobs it is especially designed to do. *The protein-sparing action of carbohydrates means that starches and sugars, by supplying energy, conserve protein for its special functions.*"

For youthful health, you need an ample amount of protein every single day. You also need carbohydrates to balance metabolism so that protein can then be conserved and spared for its youth-building functions.

A Low-Protein Diet Creates Cellular-Destruction. The U.S. Government report continues, "Suppose that the diet does not furnish enough protein for your body's daily operating and repair needs. The first thing your body does is draw on some of its own tissue protein to supply this daily wear-and-tear quota." This may create a form of *negative balance* wherein the cells can be destroyed because of a lack of sufficient nitrogen needed for nourishment. (The common proteins average 16 percent nitrogen. There must be a sufficient amount of nitrogen-bearing protein in foods, every single day. If there is a deficiency, or if more nitrogen is used up than is taken in, a negative balance occurs and tissue destruction follows.) A *positive nitrogen balance* is needed to store the protein and other materials from which to make cellular repairs. Once this is done, you will maintain a nitrogen equilibrium or balance which helps maintain the body's cellular-cerebral rejuvenation processes.

Daily Protein = Cellular Energy and Rejuvenation. To give your billions of cells the working materials for self-replenishment, daily protein intake is important. Protein will provide the energy needed for cellular rejuvenation. Basic sources are meat, poultry, fish, dairy foods, eggs, peas, beans, nuts, seeds.

CALORIES: ENERGY FOR YOUTHFUL METABOLISM

You need calories because they create energy to promote youthful metabolism, the process whereby changes take place in your body to keep your systems and networks functioning with youthful vitality.

What Is a Calorie? The answer is given in an official report by the U.S. Government[2], "A calorie is a unit of measurement that tells you how much energy you get from the foods you eat. Every food provides some of the energy that enables you to do the things you do every day. But different foods provide different amounts. To control your weight, you will need to control the amount of energy (the number of calories) you get from food, and the amount of energy you use up in exercise and normal activity.

Calories vs. Weight Control. "Whether you gain weight, lose weight, or stay the same depends on how well you balance the calories furnished by the foods you eat against the calories your body uses. If your food furnishes more calories than you use, you gain weight. If it furnishes fewer, you lose. If it furnishes just enough, your weight should stay about the same.

Guidelines for Controlled Calories and Controlled Weight. "For every 3500 extra calories you get and do not use, you gain about 1 pound of weight. This pound represents stored food energy in the form of fat. To lose excess fat, you have to somehow use up stored energy. You can—

1. Eat less food (fewer calories), to force your body to draw energy from its stored fat.
2. Increase your activity, to use up more energy.
3. Do both. Many dieters find a combination of eating less food and getting more exercise the best way to lose weight."

If you have only a few pounds to lose . . .

*Keep track of everything you eat for several days. Remember to include between-meal snacks and beverages.

---

[2]Calories and Weight, U.S. Department of Agriculture, Washington, D.C., 1974.

*Next, refer to the calorie tables in this book. Estimate the number of calories you have been getting each day. If your servings are larger than the portions given, increase the calorie counts accordingly. Mark down the amount of calories you take in daily. Then start to cut down to help lose weight.

How Much Weight to Lose? The U.S. Government report says, "Allow yourself 500 to 1000 fewer calories a day than you are now getting, to lose weight at the recommended rate of 1 to 2 pounds a week. You will need to cut down more than this on calories, however, if you are gaining weight on the amount of food you now eat. *But don't cut calories to fewer than 1200 a day unless you are under a doctor's supervision.* At calorie levels lower than this, it is difficult to get the vitamins, minerals, and proteins you need."

NINE STEPS TO CONTROLLED CALORIES

If you want to cut down on calories for weight control, the U.S. Government report offers these nine steps:

1. Take small servings and omit seconds. No matter how many calories there are in a serving of food, a smaller serving—or fewer servings—mean fewer calories.
2. Substitute lower-calorie foods for higher-calorie ones. But you do not have to completely omit your favorite high-calorie foods. Just eat them less often and have smaller servings.
3. Watch between-meal snacks. Many of the most tempting snack foods pack a lot of calories into small portions. Snacks can be part of your diet for weight control, however, if you plan for them. Be sure to include the calories they provide in your total for the day.
4. Budget your calories to allow for special occasions, such as parties. Save on calories at other meals, so you can afford extra calories for these events.
5. Each meal is important. Don't skip breakfast or lunch to cut down on calories. Skipping meals often leads to unplanned snacking. Such snacking often leads to more calories than you want—and less of some of the nutrients you need.
6. There is more to foods than calories. Follow the four Basic Food Groups (in this chapter) in making selections, to be reasonably sure of getting needed vitamins, minerals, protein, carbohydrates, and other nutrients.

7. "Crash" and "Fad" diets may be hazardous to health. Unusual foods and food combinations may seem glamorous or sure solutions to a dieter's problem, but they are not the answer.

8. Rapid weight loss is not desirable. Be satisfied to reach your target weight gradually, by making small adjustments in your eating habits.

9. Once you have reached the weight that is best for you, you will be able to eat a little more food. But continue to choose foods with an eye to caloric values, so you will not go back to the old eating habits that resulted in unwanted pounds.

HOW CALORIES PROMOTE YOUTHFUL VITALITY

Your Protein/Calories/Carbohydrates program calls for daily intake of these three essential nutrients in a proper balance. Too often, calories are drastically reduced so that cell-building protein is diverted from its rejuvenation task to be burned as energy. This task should be performed by calories which are needed daily in controlled amounts.

Calories Offer Body Energy. The official U.S. Government report on nutrition,[3] says, "The oxidation of foodstuffs by the human body provides the energy that serves it in a manner somewhat like that of a fuel in a furnace or machine.

"Your body may be thought of as a machine, which runs 24 hours a day throughout life. The machine must continue to run, even when no food is available. Then the energy comes from the oxidation of body tissue. Heat and biological energy are end products of the potential energy of the food nutrients that are oxidized within your body. All forms of energy can be converted into heat. It is convenient to express the energy exchanges of the body processes in terms of heat units or calories."

Calories Are Important. Your body requires a source of energy which is normally supplied by calories. When you eat, the food is fragmented by digestion into simple components, which are absorbed through your intestinal wall and transported to your body tissues for nourishment and energy. When these food components are oxidized within your body tissues, energy-creating calories are released. These

---

[3]*Food, Yearbook of Agriculture;* U.S. Department of Agriculture, Washington, D.C., 1959. Pages 39-56.

calories are then used to maintain your body temperature and muscle tone, to be expended in muscular work and other body activities.

Controlled Calories Maintain Health, Energy, Weight. If your food intake is *more* than enough to meet your body's energy needs, there is a storage of the excess, chiefly as body fat, and a weight gain results. If your food intake is *insufficient* to meet your body's energy needs, your body draws on its own protein-carbohydrate reserves, oxidizing them to furnish needed energy with a risk of a deficiency of these needed nutrients. Controlled calories to meet your needs will help spare the "cannibalizing" of protein and carbohydrates and your body may then have an opportunity to enjoy health, energy, and weight control.

How to Compute Your Calorie Needs. You can compute the number of calories you need each day by this simple method: Take the midpoint of the desirable weight range from the chart which follows; multiply this figure by 18 for a man, by 16 for a woman. The answer will be the approximate amount of calories needed daily by an adult of average activity.

Example: The desirable weight for a woman 5 feet 5 inches tall of average body build is about 125 pounds. If her activities are average, she will use about 125 X 16, or about 2000 calories per day.

More Activities = More Calories. If the woman is vigorously active, her calorie needs will be higher than the number of calories computed.

Less Activities = Less Calories. If the woman is relatively in-active (or an older adult), her calorie needs will be fewer than the number of calories computed. (Older folks also are likely to be less active.)

Desirable Weights

Weight in Pounds According to Frame (in Indoor Clothing)

	HEIGHT (with shoes on) 1-inch heels Feet Inches		SMALL FRAME	MEDIUM FRAME	LARGE FRAME
Men	5	2	112–120	118–129	126–141
of Ages 25	5	3	115–123	121–133	129–144
and Over	5	4	118–126	124–136	132–148
	5	5	121–129	127–139	135–152
	5	6	124–133	130–143	138–156

HEIGHT (with shoes on) 1-inch heels		SMALL FRAME	MEDIUM FRAME	LARGE FRAME
Feet	Inches			
Men	5	7	128–137	134–147	142–161
of Ages 25	5	8	132–141	138–152	147–166
and Over	5	9	136–145	142–156	151–170
	5	10	140–150	146–160	155–174
	5	11	144–154	150–165	159–179
	6	0	148–158	154–170	164–184
	6	1	152–162	158–175	168–189
	6	2	156–167	162–180	173–194
	6	3	160–171	167–185	178–199
	6	4	164–175	172–190	182–204

HEIGHT (with shoes on) 2-inch heels		SMALL FRAME	MEDIUM FRAME	LARGE FRAME
Feet	Inches			
Women	4	10	92– 98	96–107	104–119
of Ages 25	4	11	94–101	98–110	106–122
and Over	5	0	96–104	101–113	109–125
	5	1	99–107	104–116	112–128
	5	2	102–110	107–119	115–131
	5	3	105–113	110–122	118–134
	5	4	108–116	113–126	121–138
	5	5	111–119	116–130	125–142
	5	6	114–123	120–135	129–146
	5	7	118–127	124–139	133–150
	5	8	122–131	128–143	137–154
	5	9	126–135	132–147	141–158
	5	10	130–140	136–151	145–163
	5	11	134–144	140–155	149–168
	6	0	138–148	144–159	153–173

For girls between 18 and 25, subtract 1 pound for each year under 25.

CARBOHYDRATES: SOURCE OF YOUTHFUL VITALITY AND ENERGY

You need carbohydrates to give you most of the energy which you need to act and move, perform work, and live. Among the basic carbohydrates are sugars, starches, and cellulose. All green plants form carbohydrates. Here is the official U.S. Government report on carbohydrates:[4]

Carbohydrates Promote Digestive Vitality-Energy. "Carbohydrates are important in nutrition for many reasons other than as a source of energy. Some of them make our food sweet. Some determine what types of bacteria will grow in our intestines. The bulk in our food, which helps to prevent constipation, consists mostly of carbohydrates. The body needs carbohydrates in order to use fat efficiently. Some diseases, such as diabetes, develop because the body is unable to use carbohydrates properly.

Burns Fat to Control Weight. "Another important function of carbohydrate and of the series of reactions by which carbohydrate is oxidized in the body is to facilitate the oxidation of fat. *It is often said that fat is burned in a flame of carbohydrate: If not enough carbohydrate is available, it is difficult for the body to oxidize fat completely to carbon dioxide and water.* The reason is that one of the organic acids formed during carbohydrate oxidation is required for the complete oxidation of fat.

Poor Metabolism May Cause Coma. "The incomplete products of fat oxidation (caused by poor metabolism) are the short-chain, usually four-carbon organic acids known as ketone bodies. An accumulation of them can cause the blood and urine to become acidic. These acids are excreted as salts, so their excretion may lead to severe losses of sodium. This, in turn, reduces the ability of the blood to carry carbon dioxide, and in severe cases may result in coma.

Carbohydrates Help Control Weight. "This is also one of the reasons why it is unwise to try to lose weight by going without food for several days. The body's store of carbohydrate is quickly used up, and then large amounts of body fat must be used for energy. Without some carbohydrate, the fat may not be completely oxidized, and ill effects may result."

Carbohydrates + Protein = Calorie Control. The U.S. Government report continues by showing that carbohydrates plus protein can benefit with calorie control. "Carbohydrates also exert a protein-sparing effect—that is, they reduce the wastage of body or dietary

---

[4]*Food, Yearbook of Agriculture;* U.S. Department of Agriculture, Washington, D.C., 1959. Pages 88-100.

protein that occurs in certain conditions. *Calories must oxidize amino acids from protein along with fat in order to obtain energy. The result is a loss of amino acids. If carbohydrates are supplied, the body oxidizes them for energy in preference to protein, and thus the amino acids of the protein are spared for other purposes.* This sparing effect on protein is primarily an effect of calories. Calories supplied as fat can also spare protein. Carbohydrates exert a sparing effect on the protein, besides that of supplying calories and, therefore, are more effective than fat in this role."

Maintain Health of Intestines. Balanced carbohydrate intake can also help maintain health of the intestines. Carbohydrates influence the intestinal flora; that is, the bacteria and other micro-organisms that grow in the intestinal tract. They also help provide fiber or bulk. The U.S. Government report says, "Carbohydrates serve as a source of energy for the bacteria that grow in the intestine... Carbohydrates that are not absorbed quickly from the intestine also stimulate the growth of micro-organisms that synthesize many vitamins of the B-complex. Vitamins synthesized by the intestinal bacteria are absorbed and used by the body. Therefore, intestinal micro-organisms (necessary substances that are nourished by carbohydrates) are important suppliers of B vitamins.

Improves Regularity with Natural Bulk. "Carbohydrates in the intestine also help maintain normal peristaltic action—rhythmic contraction—which is favored by the presence of a certain amount of bulk in the diet. Bulk consists largely of cellulose (formed through carbohydrate metabolism) and a number of other polysaccharides (molecules) and related substances which are not digested by the enzymes of the body. They accumulate in the intestine and expand it. Since they absorb water (some of them much more than others), they contribute appreciably to the bulk of the wastes and help to prevent constipation."

Because carbohydrates have the same build-up effect in the body as do calories, they should be eaten in balance with protein. In controlled amounts, carbohydrates will help supply food energy and spare excessive use of protein so it can then help build and repair your body tissues. Carbohydrates will also help your body use fat more efficiently. In a proper balance, carbohydrates can help create good health and vitality. (See chart on next page.) Proper P/C/C planning can be your key to energetic rejuvenation and the look and feel of youthfulness.

Your P/C/C Planning Guide

Nutrient	Best Food Sources	Health Benefits	Amount Needed Daily
Protein	Lean meat, poultry, fish, sea-foods, eggs, milk, cheese, dry beans, peas, nuts, whole grain cereals, breads, fruits, vegetables, brewer's yeast, seeds.	Builds and maintains all cells-tissues. Forms an important part of enzymes, hormones, body fluids. Supplies energy.	*Adult men:* 56 grams *Adult women:* 46 grams
Calories	Almost all foods (except water), especially those high in fat, sugar, or starch.	Provide energy; help oxidize foods. Maintain body tempera-ture, muscle tone.	*Adult men:* 2400-2700 calories *Adult women:* 1800-2100 calories
Carbohydrates	*Sugars:* honey, molasses, syrups, and other sweets. Jams, jellies, candies, cakes, fruits. *Starches:* Breads, cereals, corn, grits, potatoes, rice, spaghetti, macaroni, noodles, grains, dry beans, peas, vegetables, potatoes.	Supply energy. Help body use fat efficiently. Spare protein for purposes of body building and repair. Supply bulk.	*Adult men:* 200 grams *Adult women:* 200 grams

NOTE: The amount of protein and calories required daily has been recommended by the Food and Nutrition Board of the National Academy of Science. The amount of carbohydrates needed daily has not been established but food scientists generally recommend amount listed in chart as adequate for maintenance of good health.

SIMPLE P/C/C PLANNING = EASY WEIGHT LOSS
+ YOUTHFUL VITALITY

A drastic reduction of calories and carbohydrates helped Marcia B. win the battle of the bulge. But with the loss of pounds, she also lost energy. At the end of a dieting program, Marcia B. found herself so drained of energy, she could barely do her everyday household chores. To regain her strength, she started to eat bigger meals. This made her overweight again, with restored energy. Marcia B. felt that she was in the center of a vicious and unhealthy cycle. She found a solution to this dilemma with simple Protein/Calorie/Carbohydrate planning. She needed a high-protein but low-calorie carbohydrate program. This would give her youthful vitality while her pounds melted off.

Easy P/C/C Program. Daily, Marcia B. would plan for 75 grams of protein, 1800 calories or slightly less, and 75 to 100 grams of carbohydrates. This meant that she could eat a variety of her *favorite* and "taboo" foods, provided she kept the amounts within the computerized P/C/C ratio of 75 grams of protein, 1800 calories, 75-100 grams of carbohydrates daily. This was her quick, easy weight loss program.

Benefits. The protein was used to combine with body oxygen to give Marcia B. the youthful vitality and stamina she needed while reducing. The calories offered her needed energy. The carbohydrates were used to burn fat into carbon dioxide and water so the weight could actually be "washed out" of her body. Carbohydrates were also used to *spare protein depletion* so that Marcia B. could enjoy natural energy while she was still losing weight.

Loses 50 Lbs. Looks and Feels Youthfully Alert. This easy P/C/C program helped Marcia B. lose over 50 excess pounds. But because she was nourished with sufficient protein, calories, and carbohydrates, she felt youthfully alert even while losing weight. Her appearance was bright and her attitude was cheerful. At a new slim 120, Marcia B. is now able to maintain her weight with simple computerizing of her intake of protein, calories, and carbohydrates daily. She enjoys good food, good health, good looks, good vitality . . . on a slimming program that is every bit as delicious as it is healthy!

THE EASY WAY TO PLAN A DELICIOUS DAILY DIET

Nutrition scientists have translated knowledge of the nutrient needs of people and the nutritive values of foods into an easy-to-use guide for food selection.

The *Basic Four Daily Food Guide*[5] has been prepared by the U.S. Government. It is the official plan for delicious daily dining without deprived dieting. This *Basic Four Daily Food Guide* sorts foods into four groups on the basis of their similarity in nutrient content. Each of the broad food groups has a special contribution to make toward an adequate diet. When you follow this official plan, you provide a *balance* of essential nutrients that will be used by protein, calories, and carbohydrates to keep your body in good health. These three nutrients, protein, calories, and carbohydrates *need the other nutrients* to create the cellular-cerebral rebuilding processes within your body, from head to toe. When you eat most of your favorite foods in the *Basic Four Daily Food Plan,* you will be giving your body these needed nutrients. These foods can easily fit into your pattern of eating. You enjoy good food and good health, too. (See chart below.)

A Daily Food Guide

MEAT GROUP

Foods Included

Beef; veal; lamb; pork; variety meats, such as liver, heart, kidney.
Poultry and eggs.
Fish and shellfish.
As alternates—dry beans, dry peas, lentils, nuts, peanuts, peanut butter.

Amounts Recommended

Choose 2 or more servings every day.
Count as a serving: 2 to 3 ounces of lean cooked meat, poultry, or fish—all without bone; 2 eggs; 1 cup cooked dry beans, dry peas, or lentils; 4 tablespoons peanut butter.

---

[5]*Family Fare,* U.S. Department of Agriculture, Washington, D.C., 1970. Pages 4-5.

VEGETABLE-FRUIT GROUP

Foods Included

All vegetables and fruits. This guide emphasizes those that are valuable as sources of vitamin C and vitamin A.

Sources of Vitamin C

Good sources–Grapefruit or grapefruit juice; orange or orange juice; cantaloupe; guava; mango; papaya; raw strawberries; broccoli; brussels sprouts; green pepper; sweet red pepper.

Fair sources–Honeydew melon; lemon; tangerine or tangerine juice; watermelon; asparagus tips; raw cabbage; collards; garden cress; kale; kohlrabi; mustard greens; potatoes and sweet potatoes cooked in the jacket; spinach; tomatoes or tomato juice; turnip greens.

Sources of Vitamin A

Dark-green and deep-yellow vegetables and a few fruits; namely, apricots, broccoli, cantaloupe, carrots, chard, collards, cress, kale, mango, persimmon, pumpkin, spinach, sweet potatoes, turnip greens and other dark-green leaves, winter squash.

Amounts Recommended

Choose 4 or more servings every day, including:

1 serving of a good source of vitamin C or 2 servings of a fair source.

1 serving, at least every other day, of a good source of vitamin A. If the food chosen for vitamin C is also a good source of vitamin A, the additional serving of a vitamin A food may be omitted.

The remaining 1 to 3 or more servings may be of any vegetable or fruit, including those that are valuable for vitamin C and vitamin A.

Count as 1 serving: ½ cup of vegetable or fruit; or a portion as ordinarily served, such as 1 medium apple, banana, orange, or potato, half a medium grapefruit or cantaloupe, or the juice of 1 lemon.

MILK GROUP

Foods Included

Milk—fluid whole, evaporated, skim, dry, buttermilk.
Cheese—cottage; cream; Cheddar-type, natural or process.
Ice cream.

Amounts Recommended

Some milk every day for everyone.
Recommended amounts are given below in terms of 8-ounce cups of whole fluid milk:

Children under 9   . . . 2 to 3
Children 9 to 12 . . . . 3 or more
Teen-agers . . . . 4 or more
Adults . . . . 2 or more
Pregnant women   . . . 3 or more
Nursing mothers . . . . 4 or more

Part or all of the milk may be fluid skim milk, buttermilk, evaporated milk, or dry milk.

Cheese and ice cream may replace part of the milk. The amount of either it will take to replace a given amount of milk is figured on the basis of calcium content. Common portions of cheese and of ice cream and their milk equivalents in calcium are:

1-inch cube Cheddar-type cheese   =   1/2 cup milk
1/2 cup cottage cheese   =   1/3 cup milk
2 tablespoons cream cheese   =   1 tablespoon milk
1/2 cup ice cream   =   1/4 cup milk

BREAD-CEREAL GROUP

Foods Included

All breads and cereals that are whole grain, enriched, or restored; *check labels to be sure.*

Specifically, this group includes: breads; cooked cereals; ready-to-eat cereals; cornmeal; crackers; flour;

grits; macaroni and spaghetti; noodles; rice; rolled oats; and quick breads and other baked goods if made with whole-grain or enriched flour. Bulgur and parboiled rice and wheat also may be included in this group.

Amounts Recommended

Choose 4 servings or more daily. Or, if no cereals are chosen, have an extra serving of breads or baked goods, which will make at least 5 servings from this group daily.

Count as 1 serving: 1 slice of bread; 1 ounce ready-to-eat cereal; ½ to ¾ cup cooked cereal, cornmeal, grits, macaroni, noodles, rice, or spaghetti.

OTHER FOODS

To round out meals and meet energy needs, almost everyone will use some foods not specified in the four food groups. Such foods include: unenriched, refined breads, cereals, flours; sugars; butter, margarine, other fats. These often are ingredients in a recipe or added to other foods during preparation or at the table.

Try to include some vegetable oil among the fats used.

HOW THE P/C/C DIET PLAN CREATED "NEW YOUTH" FOR "AGING" MACHINIST

Walter J. was aging rapidly. At 44, he looked 65. A co-worker at the foundry quipped that he should apply for retirement since he certainly looked the age! Walter J. felt hurt at the insult but had to admit, secretly, that the furrows in his forehead, the lines in his face, the scrawny neck, not to mention the pale, sickly complexion, all did make him look at least 20 years older. He also complained of bouts of fatigue and premature exhaustion early in the afternoon. He was tired when he awakened in the morning!

Simple Dietary Adjustments Alert His "Fountain of Youth." Walter J. made a quick and easy survey of his daily food intake. He ate many "junk foods" such as pretzels, potato chips, candy bars, soda pop. He also used huge amounts of sugar in his coffee, tea, and cooked foods. The problem here is that sugar is a "mischief maker" in the body. It supplants the need for other foods and a deficiency occurs. Walter J. was denying himself necessary foods because "junk foods" filled him up.

He made a simple adjustment. He eliminated all so-called "useless" foods. He replaced them with a high-protein, moderate-calorie carbohydrate diet plan. He scheduled himself for 100 grams of protein daily, 2500 calories daily, 75 grams of carbohydrate daily. (*Note:* As a machinist, working with heavy machines, his body used up much energy so he needed more calories and moderate carbohydrates. But he was careful to increase protein so it would build his body cells-tissues and not be crowded out by carbohydrates.) He followed the *Basic Four Daily Food Guide* (page 30) and the official U.S. Government eating plan (page 35), with easy-to-plan P/C/C foods in the amounts described above. Almost at once, he perked up. He felt as if a "fountain of youth" was gushing forth within his body.

Skin Firms Up, Schoolboy Complexion, Youthful Vitality. Walter J. saw his skin lines smooth out, wrinkles seemed to vanish, scrawny skin filled out. His complexion began to bloom and radiated with youthful vitality. So did Walter J. He now had amazing energy—bounced out of bed in the morning, refreshed, alert, and eager to face the challenges of the day. He had astonishing "go power" that kept him vital and alive throughout the hard day at the factory. Even at night, he felt youthfully alert to do chores around the house. His P/C/C program had given him a "new youth" in body and mind.

Benefits of P/C/C Diet Plan. Walter J. discovered that food can build just as so-called "food" can destroy. That is, healthy food can rebuild the body while "junk food" can destroy it. The protein in his healthy food nourished his bloodstream, rebuilt his cellular-cerebral network from head to toe, alerted his metabolism for *balanced* combustion of calories and carbohydrates. This helped make him over, give him youthful vitality. Heretofore, his "junk foods" had upset his biological clocks so that excessive sugars-starches displaced protein and created "mischief" in his body at his health's expense. Now, with this easy P/C/C Diet Plan, Walter J. could enjoy most of his favorite good foods while enjoying good health, too!

Three nutrients, *protein, calories,* and *carbohydrates,* act as "stars" in the panorama of youthful health. In proper balance, with the P/C/C diet plan, they can help rejuvenate your body and mind, and make life look and feel very beautiful!

Recommended Dietary Allowances

FOOD AND NUTRITION BOARD, NATIONAL ACADEMY OF SCIENCES-NATIONAL RESEARCH COUNCIL

RECOMMENDED DAILY DIETARY ALLOWANCES,[1] Revised 1973

Designed for the maintenance of good nutrition of practically all healthy people in the U.S.A.

	(years) From Up to	Weight (kg)	Weight (lbs)	Height (cm)	Height (in)	Energy (kcal)[2]	Protein (g)	Fat-Soluble Vitamins Vitamin A Activity (RE)[3]	(IU)	Vita-min D (IU)	Vita-min E Activity (IU)	Water-Soluble Vitamins Ascor-bic Acid (mg)	Fola-cin[6] (μg)	Nia-cin[7] (mg)	Ribo-flavin (mg)	Thia-min (mg)	Vita-min B₆ (mg)	Vita-min B₁₂ (μg)	Minerals Cal-cium (mg)	Phos-phorus (mg)	Iodine (μg)	Iron (mg)	Mag-nesium (mg)	Zinc (mg)
Infants	0.0-0.5	6	14	60	24	kg × 117	kg × 2.2	420[4]	1,400	400	4[5]	35	50	5	0.4	0.3	0.3	0.3	360	240	35	10	60	3
	0.5-1.0	9	20	71	28	kg × 108	kg × 2.0	400	2,000	400	5	35	50	8	0.6	0.5	0.4	0.3	540	400	45	15	70	5
Children	1-3	13	28	86	34	1300	23	400	2,000	400	7	40	100	9	0.8	0.7	0.6	1.0	800	800	60	15	150	10
	4-6	20	44	110	44	1800	30	500	2,500	400	9	40	200	12	1.1	0.9	0.9	1.5	800	800	80	10	200	10
	7-10	30	66	135	54	2400	36	700	3,300	400	10	40	300	16	1.2	1.2	1.2	2.0	800	800	110	10	250	10
Males	11-14	44	97	158	63	2800	44	1,000	5,000	400	12	45	400	18	1.5	1.4	1.6	3.0	1200	1200	130	18	350	15
	15-18	61	134	172	69	3000	54	1,000	5,000	400	15	45	400	20	1.8	1.5	1.8	3.0	1200	1200	150	18	400	15
	19-22	67	147	172	69	3000	52	1,000	5,000	400	15	45	400	20	1.8	1.5	2.0	3.0	800	800	140	10	350	15
	23-50	70	154	172	69	2700	56	1,000	5,000		15	45	400	18	1.6	1.4	2.0	3.0	800	800	130	10	350	15
	51+	70	154	172	69	2400	56	1,000	5,000		15	45	400	16	1.5	1.2	2.0	3.0	800	800	110	10	350	15
Females	11-14	44	97	155	62	2400	44	800	4,000	400	10	45	400	16	1.3	1.2	1.6	3.0	1200	1200	115	18	300	15
	15-18	54	119	162	65	2100	48	800	4,000	400	11	45	400	14	1.4	1.1	2.0	3.0	1200	1200	115	18	300	15
	19-22	58	128	162	65	2100	46	800	4,000		12	45	400	14	1.4	1.1	2.0	3.0	800	800	100	18	300	15
	23-50	58	128	162	65	2000	46	800	4,000		12	45	400	13	1.2	1.0	2.0	3.0	800	800	100	18	300	15
	51+	58	128	162	65	1800	46	800	4,000		12	45	400	12	1.1	1.0	2.0	3.0	800	800	80	10	300	15
Pregnant						+300	+30	1,000	5,000	400	15	60	800	+2	+0.3	+0.3	2.5	4.0	1200	1200	125	18+[8]	450	20
Lactating						+500	+20	1,200	6,000	400	15	60	600	+4	+0.5	+0.3	2.5	4.0	1200	1200	150	18	450	25

[1] The allowances are intended to provide for individual variations among most normal persons as they live in the United States under usual environmental stresses. Diets should be based on a variety of common foods in order to provide other nutrients for which human requirements have been less well defined.

[2] Kilojoules (KJ) = 4.2 × kcal

[3] Retinol equivalents

[4] Assumed to be all as retinol in milk during the first six months of life. All subsequent intakes are assumed to be one-half as retinol and one-half as β-carotene when calculated from international units. As retinol equivalents, three-fourths are as retinol and one-fourth as β-carotene.

[5] Total vitamin E activity, estimated to be 80 percent as α-tocopherol and 20 percent other tocopherols.

[6] The folacin allowances refer to dietary sources as determined by Lactobacillus casei assay. Pure forms of folacin may be effective in doses less than one-fourth of the RDA.

[7] Although allowances are expressed as niacin, it is recognized that on the average 1 mg of niacin is derived from each 60 mg of dietary tryptophan.

[8] This increased requirement cannot be met by ordinary diets; therefore, the use of supplemental iron is recommended.

Highlights

1. The P/C/C program calls for balanced daily intake of three nutrients that help ignite the "fountain of youth" within your own body.

2. Protein is the prime nutrient that acts as a "staff of life." The official U.S. Government report recommends daily intake of this essential life-giving, youth-restoring nutrient.

3. Calories help provide energy for youthful metabolism. The official U.S. Government report tells how to use calories for controlled weight.

4. Carbohydrates are a little-known but highly important source of youthful vitality and energy. The official U.S. Government report tells how this nutrient can help promote youthful digestive vitality and healthy energy.

5. Plan new health with your P/C/C guide which shows you the best food sources for these three nutrients, their health benefits, and amount needed daily. Follow this plan with the protein/calorie/carbohydrate counts in this book for better health and youthful vitality.

6. Marcia B. enjoyed easy weight loss plus youthful vitality with simple P/C/C planning.

7. Walter J. discovered "new youth" as the P/C/C diet plan, based on the U.S. Government's official *Basic Four Daily Food Guide,* helped turn back the aging clock. His easy P/C/C plan firmed up his skin, gave him healthy energy and a new lease on life. It can help you, too!

2/ How to Use the Protein/Calorie/ Carbohydrate Computer for Youthful Health

It's easy to plan your daily intake of protein, calories, and carbohydrates for youthful health with a simple chart that you can compute right in your own home, using only a pencil and a piece of paper. When you prepare your own P/C/C chart, you then select any of the thousands of delicious foods listed in the charts in this book, according to the desired counts and measurements to fit within your needs. When you fortify your body with needed protein, together with required calories and carbohydrates in the amounts needed by your body, you then give your cellular-cerebral-metabolic systems the working materials out of which can be created the elements that help create youthful health . . . from head to toe. Together with a healthful eating program, these three nutrients can protect against illness, set off the biological clocks within your metabolism to tick in a youthful rhythm, and give you the look and feel of youth.

HOW TO COMPUTE YOUR "HIGH" OR "LOW"
PROTEIN/CALORIE/CARBOHYDRATE REQUIREMENTS

For good health, your computerized P/C/C chart should tell you whether you need high or low protein, calories, or carbohydrates in your daily eating program. You decide on your specific needs before you compute your P/C/C chart by deciding on the health improvements you want. Here is how to do it.

"High" vs. "Low" Daily Protein Needs

High Protein. Necessary for improved growth, regulation of body processes such as production of hormones, improvement of bloodstream, rebuilding of cells and tissues, rejuvenation of skin and hair. Also needed to create more youthful energy if you are on a low-fat, low-carbohydrate program, since these two nutrients influence your energy rates.

Low Protein. Presence of ketones or waste substances in the urine may be caused by an excess of protein and a deficiency or absence of carbohydrates. Problems of dehydration are caused when very high protein and low or no carbohydrates create imbalance and there is too much water loss as well as sodium-potassium depletion. By maintaining low protein intake, body metabolism can be restored to equilibrium and better balance.

"High" vs. "Low" Daily Calorie Needs

High Calories. Help correct problems of severe underweight, loss of energy, chronic fatigue, sluggish metabolism, poor circulation, feeling of chill and cold. When poor appetite has created underweight, a higher calorie intake will help stabilize weight and improve general health.

Low Calories. If you want to lose weight, you need a reduced calorie program "computerized" to show that 3500 fewer calories help you lose 1 pound. For 2 pounds loss a week, you "computerize" your plan to show 7000 fewer calories during that week. You need a healthy amount of calories per week, depending upon your height and body build, as described in Chapter 1. Computerize your needs, then plan your daily calorie counts, based upon the thousands of foods in this book.

"High" vs. "Low" Daily Carbohydrate Needs

High Carbohydrates. Necessary when you feel weak, listless, low in energy, sluggish in your various body processes, when you have an accumulation of fat that cannot be metabolized; when you want protein to be spared for more essential processes such as rebuilding of your body cells and tissues. When you need quick energy, your body can speedily break down carbohydrate to utilize its glucose, a substance that provides youthful vitality.

Low Carbohydrates. Useful when you want to lose weight and have an adequate or higher protein intake so you will be energetic even when shedding pounds. Also helps correct problems of hypoglycemia, or a low level of blood sugar. Carbohydrate starch is a major source of hidden sugar which creates mischief with the blood sugar levels. That is, excess carbohydrate starch-created sugar can upset the delicate levels to provide bursts of energy and then a serious mental-physical letdown. A low carbohydrate food program, balanced by high protein and moderate calories, will help maintain healthy levels of blood sugar and protect against hypoglycemia. Low carbohydrate intake with high protein also helps control weight.

HOW TO PREPARE A P/C/C COMPUTER CHART—FOR MEN

Fred A. had recurring backaches, annoying headaches, accompanied by frequent dizzy spells. He found it difficult to start walking as soon as he got up out of a chair. He had to hold on to something (or someone) for a few moments to regain his balance. Fred A. felt he was prematurely aging. He looked aged, with deep crease lines, skin blotches, sagging throat muscles. When someone remarked that he looked like his wife's father, he felt impelled to take steps to correct the embarrassing problem. He began with creating his own P/C/C Computer Chart. It was designed for men only. (See page 40.)

(*Note:* Because the protein and calorie needs are different for men and women, charts should be prepared for individual use by the specific sex.)

High Protein, Low Calorie, High Carbohydrate. Fred A. created his P/C/C Chart with emphasis upon protein since he needed this nutrient in higher amounts to help nourish the billions of his tissues that were breaking down because of a deficiency. Protein would also protect against a negative nitrogen balance, a problem that may lead to destruction of body cells. This was one probable cause of his backaches as well as his dizzy spells, not to mention recurring headaches.

Fred A. reduced his calorie intake to help take off unnecessary pounds. Previously, he had eaten many "junk foods" with so much refined sugar intake, he had gained at least 40 heavy pounds. With a low calorie program, the pounds started to melt away. Now he felt lighter on his feet, was able to get up from his chair or bed and walk with a bounce and a youthful lilt.

Fred A. increased his carbohydrate intake. The benefit here is that carbohydrate would spare protein, protecting against protein wastage in the body. At the same time, carbohydrates helped to oxidize accumulated fat in his body, transforming it to carbon dioxide and water. This created better metabolism so that his entire

SAMPLE CHART

For Adult Males:

(As used by Fred A.)

Protein/Calorie/Carbohydrate Computer

	Protein Grams	Calories	Carbohydrates Grams
Breakfast	50	750	75
Luncheon	50	750	125
Dinner	100	900	150
Total Daily Intake	200	2400	350

Required Daily Intake for Youthful Health:

Protein—200 grams daily
Calories—2400 calories daily
Carbohydrates—350 grams daily

body, from head to toe, felt slimmer, cleaner, lighter, and more youthful. It was the easy way to lose weight and restore health.

This high-protein, low-calorie, high-carbohydrate plan erased his problems, tightened up his skin, made him look like his wife's younger brother, instead!

HOW TO COMPUTE YOUR P/C/C DAILY INTAKE (MALES)

1. To compute your needs, (as Fred A. did,) total up your daily counts of protein, calories, and carbohydrates from foods as listed in this book.

2. Compare the total counts you have eaten with the required daily intake for youthful health.

3. Either subtract or add and then calculate how much more (or less) protein, calories, and carbohydrates you need daily. When you have reached a satisfactory daily count, try to keep it at that level for youthful health.

HOW TO PREPARE A P/C/C COMPUTER
CHART—FOR WOMEN

Lena K. had a difficult "middle years" problem. She felt increasing tension, hot flashes, moist palms, cold feet. Frequently, Lena K. was given to emotional outbursts of temper. Unable to sleep adequately, she would get up in the middle of the night and start eating whatever she could find. This created overweight. She was difficult to live with. Her family and friends tried to be sympathetic, but her unstable behavior was so unpredictable that she alienated herself from her loved ones. A sympathetic friend told her that menopausal-like symptoms could be eased with a proper balance of protein, carbohydrates, and calories. A dietary correction would help adjust her hormonal system so that she could cope with the "middle years" and ease up on the symptoms. So it was that Lena K. prepared her own P/C/C Computer Chart. It was designed for women only. (See page 43.)

High Protein, Low Calories, Low Carbohydrates. The P/C/C Chart prepared by Lena K. called for a higher intake of daily protein. This nutrient would nourish her female glands to respond to her "change of life" reaction. Protein was needed to help level off the production of estrogen, the female hormone, with a gradual rhythm so there would be a minimum of reaction. Protein was also needed to nourish the hemoglobin (red coloring matter of the bloodstream) so that it would be able to transport oxygen from the lungs to the body tissues. This helped create better respiration so there was a natural warming of the hands and feet. Better oxidation also eased excessive perspiration. Protein also took the body's supply of B-complex vitamins to soothe the nervous system so that Lena K. could relax, find release from her tensions, and be able to enjoy a good night's sleep. Also, a higher protein intake meant more polyunsaturated fatty acids; that is, the healthy type of fat that offered a feeling of appetite satisfaction. Automatically, this controlled her appetite. She stopped overeating. Slowly, she became more emotionally stabilized under a higher-protein intake.

Low Calories Help Her Reduce. Because Lena K. was less active physically than other women, she required lower calories. She planned her P/C/C Chart to include a minimal amount of calories. The benefit here is that reduction of calories meant reduction of weight. She shed at least 4 pounds per week, without feeling hungry (high protein satisfied her appetite). Lena K. also felt more energetic under a controlled calorie intake. The reason here is that her digestive system was less burdened under moderate calories than under heavy calories. When she ate wholesome food, the components were oxidized within her body cells and tissues to release calories that would create needed energy. They also helped regulate her body temperature (she said "goodbye" to hot flashes and sweats), and also helped give her smooth and youthful muscle tone. As she slimmed down, she felt lighter on her feet, more energetic, more youthful. Life became worth living again.

Low Carbohydrates for Better Vitality. Previously, Lena K. had a sluggish metabolism which meant that her billions of body cells were clogged with fats and wastes. A low carbohydrate plan meant that this nutrient could burn up the fat, help oxidize it, and then wash it out of the system. When Lena K. had overeaten, the excessive calories blocked the fat-melting action of the carbohydrates. Also, excessive carbohydrates meant that accumulated clumps of these sugars and starches could not efficiently melt away body fat. The reaction can be compared with pouring liquids through a funnel into a slender-necked jar or bottle. If you overflow the pouring funnel, the liquid spills over the side "choking" the funnel, with the result that little will enter the bottle. But if you use a slow and modest supply of liquid, it can flow smoothly through the funnel, into the bottle. So it is with carbohydrate intake. An excessive amount "chokes" the cells so that there is reduction of fat oxidation. A modest or low carbohydrate intake enables the burning process to continue at a healthy and successful rate of speed. So it was that Lena K. was able to stabilize her hormone balance, lose excessive weight, and calm down her nervous system with a properly balanced P/C/C Computer Chart.

Life was very young and very beautiful again for Lena K., on this easy-to-follow program.

SAMPLE CHART

For Adult Females:			
(As used by Lena K.)			
Protein/Calorie/Carbohydrate Computer			
	Protein Grams	Calories	Carbohydrates Grams
Breakfast	50	300	30
Luncheon	45	400	20
Dinner	80	800	50
Total Daily Intake	175	1500	100

Required Daily Intake for Youthful Health:

Protein—175 grams daily
Calories—1500 daily
Carbohydrates—100 daily

HOW TO COMPUTE YOUR P/C/C DAILY INTAKE (FEMALES)

1. To compute your needs (as did Lena K.), total up your daily counts of protein, calories, and carbohydrates from foods as listed in this book.

2. Compare the total counts you have eaten with the required daily intake for youthful health.

3. Either subtract or add and then calculate how much more (or less) protein, calories, and carbohydrates you need daily. When you have reached a satisfactory daily count, try to keep it at that level for youthful health.

P/C/C PLAN + BASIC NUTRIENTS = YOUTHFUL HEALTH

For your body to be rewarded with most of the benefits of proper protein/calorie/carbohydrate planning, it needs to be adequately nourished with the other basic nutrients which will add up to good health, brimming with youth. Do not sacrifice these nutrients. Your body needs a *balance* of all nutrients with controlled amounts of protein/calorie/carbohydrates for efficient metabolism and youthful health. Daily, you should nourish your body with:

Vitamin A

Benefits. Used by P/C/C to help eyes adjust to dim light, keep skin smooth, keep lining of mouth, nose, throat, and digestive tract healthy and resistant to infection, to promote growth.

Food Sources. Liver, dark-green and deep-yellow vegetables (such as broccoli, turnip, and other leafy greens), carrots, pumpkin, sweet potatoes, winter squash, apricots, cantaloupe, butter, fortified margarine.

Vitamin B-Complex

Benefits. Used by P/C/C to help your body cells obtain energy from food, to keep nerves in healthy condition, to promote good appetite and digestion, to help cells of body use oxygen to release energy, to keep eyes, skin healthy.

Food Sources. Lean meat, heart, kidney, liver, dry beans, peas, whole grain breads and cereals, seeds, nuts, eggs, milk, dark leafy greens, peanuts and peanut butter, brewer's yeast.

Vitamin C

Benefits. Used by P/C/C to protect against infection, promote sound teeth and gums, help distribute and diffuse calcium for better tooth and bone formation, prevents scurvy, protects against bron-chial-respiratory disorders, and also against infections of colds.

Food Sources. Tomatoes, all citrus fruits (lemons, grapefruit, oranges, etc.).

Vitamin D

Benefits. Used by P/C/C to promote absorption and utilization of calcium and phosphorus, essential for normal bone and tooth development.

Food Sources. Fish oils, egg yolk, liver, Vitamin D-fortified milk. Direct sunlight produces Vitamin D from cholesterol in your skin.

Vitamin E

Benefits. Used by P/C/C to act as an antioxidant to keep metabolic processes in balance. Helps prolong life of red blood cells. Prevents formation of toxic substances. P/C/C uses this vitamin to protect Vitamin A and other easily oxidizable vitamins. Helps protect against aging process.

Food Sources. Wheat germ oil, whole grain breads and cereals, leafy vegetables, peanuts, walnuts, liver.

Calcium

Benefits. Used by P/C/C to build bones and teeth, help blood to clot, help nerves, muscles, and heart function properly, maintain proper heartbeat, convert chemical energy into muscle contractions, influence neuromuscular irritability control.

Food Sources. Milk (whole, skim, buttermilk, yogurt), cheese, leafy vegetables such as collards, dandelion, kale, mustard greens, turnip greens, salmon, cabbage, apricots.

Phosphorus

Benefits. Used by P/C/C to unite with calcium to form and maintain bones and teeth in good health and to regulate a healthful acid-alkaline balance. P/C/C uses this mineral to transform biological energy into physical reactions to help promote a healthy metabolism and a feeling of youthful energy.

Food Sources. Liver, kidney, heart, lean meat, egg yolks, dry beans, peas, cottage cheese, broccoli, whole grain breads and cereals, soybeans, bran, fish, poultry, brewer's yeast.

Iron

Benefits. Used by P/C/C to combine with other nutrients to make hemoglobin, the red substance of blood. P/C/C then uses iron to carry oxygen from the lungs to the muscles, brain, and other parts of the body. P/C/C uses iron to help the cells metabolize oxygen.

Food Sources. Liver, kidney, heart, lean meat, egg yolks, dry beans, dark-green leafy vegetables, sun-dried fruits, whole grain breads and cereals, molasses.

Copper

Benefits. Used by P/C/C to promote synthesis of red blood cells and oxidation systems of the body. P/C/C uses this mineral in combination with iron for the formation of hemoglobin. A good blood-building mineral.

Food Sources. Liver, kidney, fish, nuts, seeds, sun-dried fruits such as raisins, dried beans, peas, soybeans.

P/C/C BALANCE CREATES BODY-MIND BALANCE

A simply computerized protein/calorie/carbohydrate intake is the foundation for the creation of a balanced health reaction of the body and mind. P/C/C works with all vitamins and minerals to create the biological reactions that help promote youthful health of your body and mind.

High Protein Plan Promotes Blood Building. Evelyn S. was always cold. Her hands and feet were chilled even in the middle of a heat wave! She had a pale, almost sickly complexion. Her resistance was so low that she caught colds throughout the year. She felt weak after very little exertion. Evelyn S. could not understand what was wrong. She ate many iron-containing foods, assuming that this blood-building nutrient would correct her obvious deficiency. But her error was in focusing attention solely on one nutrient. She soon learned that a P/C/C program would correct her deficiency since a high-protein intake meant that the iron could then be metabolized more efficiently.

High-Protein/Low-Calorie/Low-Carbohydrate Plan Enriches Her Bloodstream. Whenever Evelyn S. had an iron-containing food, she balanced it with a high-protein food, also she followed a low-calorie and low-carbohydrate plan. The reason here is that food iron is not readily absorbed because it is present in organic complexes which must first be broken down to free the mineral. It is broken down by protein which, in turn, is energized by smaller amounts of calories and carbohydrates. (A high calorie-carbohydrate diet overburdens the digestive system and creates a sluggish iron-splitting protein action.)

Evelyn S. boosted her intake of lean meat, poultry, fish, skim milk, dry beans, peas, seeds, and nuts. She cut down on her sugars and starches. Simultaneously, she ate more iron-containing foods at the same meal with her high-protein foods.

Benefits. The protein could not break down the organic complexes of the iron and transform it into a form that could then be assimilated by the bloodstream. In a short while, she felt herself glow with youthful warmth. Her hands and feet felt warm. Her complexion looked as if she had roses in her cheeks! She resisted cold and bronchial disturbances throughout the week. She now felt energetic and alert. So it was that she learned that iron, of itself, is largely inactive without the availability of balanced protein/calorie/carbohydrate. A simple adjustment made her bloom again with the glow of radiant youth, thanks to P/C/C.

With protein/calories/carbohydrates in proper amounts, based upon your needs, combined with balanced nutrition, your body will respond with vitality and energy so you will look and feel young all over.

In Review

1. Compute your "high" or "low" protein/calorie/carbohydrate requirements for better health, using the sample charts in this chapter as guidelines.
2. Fred A. corrected internal imbalance, soothed his recurring backaches, headaches, and dizzy spells, as well as premature aging, with a P/C/C Computer Chart he created in his own home.
3. Lena K. breezed through the "middle year" change of life with balanced nutrition and her P/C/C Chart.

4. Your daily food plan should include the P/C/C ratio together with other basic nutrients as listed for youthful health of body and mind.

5. Evelyn S. boosted circulation and warmed up her cold hands and feet, built resistance to respiratory infections, and bounced back to health with joyful energy when she used a high-protein plan with her iron intake.

3 /

How to Read Food Labels (Including Weights, Measures, Metric System) for Your P/C/C Planning Program

Want to know what nutrients are in the food you eat? It's simple. Just read the label on the package. The Food and Drug Administration has developed a labeling program to help you identify the nutrient content of the foods you buy. All labels with nutrition information must follow the same format. Any food to which a nutrient is added, or which makes a nutritional claim, must have a nutritional label. Nutrition labeling for other foods is optional.

Nutrients Listed in Order. In addition to the usual information, such as name of the food manufacturer, net weight, and ingredients, the label tells you what nutritional value is in the food, with nutrients listed in order:

> Calories
> Protein
> Carbohydrate
> Fat
> Vitamin A
> Vitamin C
> Thiamine
> Riboflavin
> Niacin
> Calcium
> Iron

Note: These labels are examples of what you should look for in your store when purchasing a packaged product. The label tells you how much protein, calories, and carbohydrates are found in the product, per serving. This makes it easy for you to plan your daily P/C/C program.

NUTRITION INFORMATION
(PER SERVING)
SERVING SIZE = 1 OZ.
SERVINGS PER CONTAINER = 12

Calories	110 grams
Protein	2 grams
Carbohydrate	24 grams
Fat	0 gram

PERCENTAGE OF U.S. RECOMMENDED DAILY
ALLOWANCES (U.S. RDA)*

Protein	2
Thiamine	8
Niacin	2

*Contains less than 2 percent of U.S. RDA for Vitamin A, Vitamin C, Riboflavin, Calcium, and Iron.

This is the minimum information that must appear on a nutrition label.

NUTRITION INFORMATION
(PER SERVING)
SERVING SIZE = 8 OZ.
SERVINGS PER CONTAINER = 1

Calories 560	Fat (Percent of	
Protein .... 23 g	Calories 53%)	33 g
Carbohydrate .... 43 g	Polyunsaturated	2 g
	Saturated	9 g
	Cholesterol*	
	(20 mg/100 g)	40 mg
	Sodium (365 mg/	
	100 g)	830 mg

PERCENTAGE OF U.S. RECOMMENDED DAILY
ALLOWANCES (U.S. RDA)

Protein 35	Riboflavin	15
Vitamin A 35	Niacin	25
Vitamin C	Calcium	2
(Ascorbic Acid) 10	Iron	25
Thiamine (Vitamin B$_1$) 15		

*Information on fat and cholesterol content is provided for individuals who, on the advice of a physician, are modifying their total dietary intake of fat and cholesterol.

A label may include optional listings for cholesterol, fats, and sodium.

Standard Format. Nutrition information must appear in a standard format. Unless space does not permit, the information must always appear on the part of the label immediately to the right of the main panel. Having the nutrients always listed in the same order and location makes it easier for you to compare labels.

P/C/C Information on Label. The *upper* portion of the label (see page 52) shows you the amount of protein and the number of calories and carbohydrates in a serving of the food. The *lower* portion of the label tells you the percentage of United States Recommended Daily Allowance (U.S. RDA) (see page 52 and also 53) for these three nutrients, P/C/C, as well as seven vitamins and minerals and fat provided in one serving. You can easily compute your daily needs for protein, calories, carbohydrates, and these other nutrients. Then just add up how much is in each serving of each food you prepare and you should be getting your desired daily amount.

Nutrients listed are one serving

NUTRITION INFORMATION
(per serving)

Serving Size = 1 cup
Servings per Container = 2

Nutrients in metric weight as grams (1 ounce = 28 grams)

Number of servings per container

Calories 110
Protein . . . . 1 gram
Carbohydrate . . . . . . 25 grams
Fat . . . . 1 gram
Sodium
    (970mg/100gm) . . 275 milligrams

Percentage of U.S. Recommended Daily Allowances (U.S. RDA)

Percentages of U.S. Recommended Daily Allowances

Labels may show amounts of cholesterol and sodium in 100 grams of food and in a serving

Protein 2
Vitamin A 25
Vitamin C 25
Thiamine 25
Riboflavin 25
Niacin 25
Calcium 4
Iron 4

U.S. RDA

The U.S. RDAs are the amounts of protein, vitamins, and minerals people need each day to stay healthy.

These allowances are set by the Food and Drug Administration. They are based on body needs for most healthy adults.

Set at generous levels, they provide a considerable margin of safety for most people above minimum body needs for most nutrients.

Nutrition labels list U.S. RDAs *by percentage* per serving of food.

U.S. RECOMMENDED DAILY ALLOWANCES (U.S. RDA) FOR ADULTS AND CHILDREN OVER 4 YEARS OLD

Nutrients	*Amounts*
Protein	45 or 65 grams*
Vitamin A	5000 International Units

* 45 grams if protein quality is equal to or greater than milk protein, 65 grams if protein quality is less than milk protein.

Nutrients	Amounts
Vitamin C (ascorbic acid)	60 milligrams
Thiamine (vitamin B_1)	1.5 milligrams
Riboflavin (vitamin B_2)	1.7 milligrams
Niacin	20 milligrams
Calcium	1.0 gram
Iron	18 milligrams
Vitamin D	400 International Units
Vitamin E	30 International Units
Vitamin B_6	2.0 milligrams
Folic acid (folacin)	0.4 milligram
Vitamin B_{12}	6 micrograms
Phosphorus	1.0 gram
Iodine	150 micrograms
Magnesium	400 milligrams
Zinc	15 milligrams
Copper	2 milligrams
Biotin	0.3 milligram
Pantothenic acid	10 milligrams

HOW TO READ NUTRITION LABELS

Nutrition information is per serving. The label gives the size of a serving (for example, 1 cup, 2 ounces, 1 tablespoon) and tells how many servings are in the container.

Then calories are listed, followed by the amounts in grams of protein, carbohydrate, and fat.

Protein is listed twice, in grams and as a percentage of the U.S. Recommended Daily Allowance.

Seven vitamins and minerals must be shown, in the same order, on all nutrition labels.

Other vitamins and minerals may also be listed.

Listing of cholesterol, fatty acid, and sodium content is optional.

HOW TO USE NUTRITION LABELING FOR P/C/C PLANNING

1. Compare labels to select foods that round out the amount of protein, calories, and carbohydrates you need, together with other nutrients. *For example,* if you need more protein, compare food labels to find the best sources of this nutrient.

2. Use nutrition labels to help you count calories and carbohydrates.

3. If you are on a special diet recommended by your physician,
use nutrition labels to help avoid restricted foods.

4. Read labels on *new* foods to see what nutrients they supply.

When you read labels, you can plan for balanced intake of
protein, carbohydrates, and calories based upon your personal
requirements.

WHAT THIS SAMPLE LABEL TELLS YOU
ABOUT P/C/C PLANNING

NUTRITION INFORMATION

Per Serving

Serving Size	1 cup
Servings per Container	3½
Caloric Content (Serving)	190
Protein Content	1 gram
Carbohydrate Content	50 grams
Fat Content	0 gram

Percentage of U.S. Recommended Daily Allowances
of Protein, Vitamins, and Minerals (Serving):

Protein	*
Vitamin A	20
Vitamin C	15
Thiamine	*
Riboflavin	2
Niacin	8
Calcium	*
Iron	2
Phosphorus	2
Magnesium	2

*Contains less than 2 percent of the U.S. RDA of these nutrients.

At a glance, you can see that a 1-cup serving of this food will
give you 190 calories and 50 carbohydrate grams but only 1 protein
gram. This product is good if you are on a moderate-calorie and
higher-carbohydrate program with low-protein intake. You would

need to balance it with other foods to get more protein, if that is your plan. Just read the labels for guidance and easy planning.

"Fortified" or "Enriched" Products. These products must, by law, carry nutrition labeling. "Fortified" describes the addition of one or more nutrients which are naturally present in a lesser amount. "Enriched" describes the addition of one or more nutrients which are not naturally present (to increase food value).

Plan Your P/C/C Program. Begin by deciding how much protein, calories, and carbohydrates you require daily, along with any other nutrients. Keep a daily record of everything you eat. Tabulate the percentages of nutrients from the nutrition label. At the end of the day, add up the totals. Compare it with your P/C/C chart. Evaluate your results by comparing how close you come to your needs. You may have to make some adjustments in your meal planning until you come close or exactly to the P/C/C counts required.

Compare Nutritional Values. Use nutrition labeling information in the store to choose the best nutritional buys. Compare "house brands" and the nationally advertised brands to see what nutrition you are getting for your money. Quite often, you can get more nutrition for your dollar with house brands.

YOUR GUIDE TO WEIGHTS AND MEASURES

You can easily compute your daily intake of protein, calories, and carbohydrates by using the following chart of weights and measures. When preparing a recipe or planning a menu with prepared foods, just follow these measurements to compute your required amounts of protein, calories, and carbohydrates:

A few grains, pinch, dash, etc. (dry)	= less than 1/8 teaspoon
1 tablespoon	= 1/2 ounce
	= 3 teaspoons
2 tablespoons	= 1 ounce
	= 1/8 cup
4 tablespoons	= 2 ounces
	= 1/4 cup
5 tablespoons + 1 teaspoon	= 1/3 cup
8 tablespoons	= 1/2 cup
12 tablespoons	= 3/4 cup
16 tablespoons	= 1 cup

1 cup	= 8 fluid ounces
	= 1/2 pint
2 cups	= 1 fluid pint
4 cups	= 1 fluid quart
1 quart	= 2 pints
4 quarts	= 1/2 gallon
1/4 pound	= 4 ounces
1/2 pound	= 8 ounces
1 pound	= 16 ounces

YOUR KEY TO METRIC UNITS

Nutrition labels often show amounts in grams rather than ounces, because grams are a smaller unit of measurement. Many food components are present in small amounts. The metric system is used on many labels, especially products of foreign origin. Here is a guide to help you read metric units on labels:

1 milligram (mg.)	= 1000 micrograms (mcg.)
1 gram (gm.)	= 1000 milligrams, 15.43 grains, .035 ounces
1 ounce	= 28.35 grams
1 pound	= 453.59 grams

OTHER EQUIVALENTS

1 microgram	= 1 gamma
1 ounce	= 480 minims, 29.57 cubic centimeters (cc.)
1 kilogram	= 2.2046 pounds, 1000 grams
1 cc	= 16.23 minims
1 teaspoon	= 62.5 minims, 5 cc
1 tablespoon	= 15 cc, 1/2 ounce, 14.18 grams
1 cup	= 8 ounces, 226.78 grams

TERMS

I.U.	= International Unit (1/20 of 1 mg.)
MDR	= Minimum Daily Requirement
USP Unit	= United States Pharmacopoeia Unit
1 USP Unit Vitamin A	= 0.6 mcg.
1 USP Unit Vitamin B_1	= 0.003 mg.
333 USP Unit Vitamin B_{12}	= 1.0 mg.
1 USP Unit Vitamin C	= 0.05 mg.
1 USP Unit Vitamin D	= 0.0000025 mg.
1 USP Unit Vitamin E	= 1.0 mg.

HOW TO READ UNIT PRICING LABELS

Here are the unit price measures you'll see—

- **Items sold by WEIGHTS and MEASURES:**
  Per ounce, per pound, per quart.
  Most grocery items, frozen food.

- **Items sold by COUNT:**
  Number on a package such as napkins, paper towels, paper bags.

- **Items sold by AREA:**
  Square feet such as foil, food wrap.

  REMEMBER, compare only *like* items to determine your best buy. Don't compare raspberry jam with apple jelly, for example. Compare ounces with ounces, quarts with quarts for accurate information. Because there are so many different items from household products, Health and Beauty Aids to food in liquid and solid state, the same unit of measure cannot be used for all.

  Price, of course, is only a yardstick. You also should consider: *quality, taste, personal preference, convenience,* and *size.* (See sample unit pricing label on page 58.)

To save money:

One: Use labels to compare the cost per serving of similar foods.

Yields: 4 (½-cup)* servings	Yields: 7 (½-cup)* servings
(31¢ ÷ 4 = Cost)	(49¢ ÷ 7 = Cost)
Cost 7.8¢ per serving	Cost: 7¢ per serving

Brand X 31¢ 16 oz.	Brand Y 49¢ 29 oz.

*Note: Serving sizes must be the same for accurate comparison.

A new kind of shelf tag lets you make accurate price comparisons at a glance!

Comparison measure

Price per unit use to compare prices

The Unit Price is always on the left hand side

The retail price you pay

Brand and name of item

Weight of item

Number of packages per case

Commodity code

UNIT PRICE
42.5¢
PER
LB 00

YOU PAY
85¢
GRAPE JELLY 2 LB
21690 -12

The larger size—
2 lb. jar is the best buy.

UNIT PRICE
65.3¢
PER
LB 00

YOU PAY
49¢
GRAPE JELLY 12 OZ
21670 -24

UNIT PRICE
42.5¢
PER
LB 00

YOU PAY
85¢
GRAPE JELLY 2 LB
21690 -12

Compare on Value
To compare costs of a product in different sizes or brands, just look at the price on the unit price side of the label . . . then compare that unit price against another, checking price per measure, brand to brand, size to size.

Two: Read labels to make sure you get the most for your food dollar. For example, compare two frozen pot pies of the same weight. One costs 39 cents, the other 29 cents. But when you read the nutrition label, you may see the pot pie that costs 39 cents provides a higher percentage of the U.S. RDA for protein. So if you are serving the pot pie as a main dish, and protein content is important, the one that costs 39 cents may be a better buy—nutrition-wise.

Three: Read labels to find less-costly substitutes for more expensive foods. For instance, you may be surprised to learn that many canned and packaged foods have high amounts of protein at a reasonable price.

HOW LABEL READING CREATED A P/C/C PLAN FOR HEALTHY WEIGHT LOSS

Paul L. was more than 70 pounds overweight. Walking short distances made him huff and puff. His face grew florid upon the slightest exertion. He had a 50-inch waistline that made him the brunt of jokes. His mental anguish was as upset as his physical health. His overweight made him feel dizzy. There were recurring bouts of colitis because of his overeating, compounded with feelings of nervous tension and elevated blood pressure. Paul L. wanted to reduce, but in so doing, he cut down on his protein foods because many protein foods are also high-calorie foods. He wanted to maintain his daily intake of about 150 grams of protein but wanted to restrict his calorie intake to about 1400 per day. He also wanted to lower his carbohydrate intake to less than 50 grams daily. Again, Paul L. faced a problem wherein his favorite foods were high in protein, but also high in calories and carbohydrates. So he kept gaining weight—until he started label reading.

High-Protein, Low-Carbo-Cal Plan Offers Speedy Weight Loss. On all packaged foods, he read the labels carefully. Thus, he could select those foods which were high in protein but very low in carbohydrates and calories. This included fish, very lean meat products, packaged dairy products made from skim milk, special types of whole grain products, to name a few. On this label-reading program, Paul L. could plan for 150 grams of protein daily, with 1400 calories and 50 grams as the absolute maximum. Soon, his 70 pounds started to melt away. His waist slimmed to a youthful 36, and still kept reducing. Once the heavy weight was lost, Paul L. could walk up and down stairs with ease, his face no longer was flushed and florid, he conquered feelings of dizziness as well as the problem of colitis. Regularity was established when he slimmed to youthful health. His blood pressure was soothing and comfortable. He is able to keep slim and healthy on his P/C/C plan by simple label reading.

EASY SUPERMARKET CALCULATING
FOR BETTER SHOPPING*

In childhood you were surely told that a knowledge of math will help you make everyday decisions. But how often do you feel that some arithmetic might help but you don't know what kind to use? Even if you know what to do, how often are the actual mechanics just too cumbersome?

Say you want to quickly check a short grocery bill, but the difficulty of accurately adding a column of 29s, 38s, and 59s in your head makes it too much trouble.

The Near Numbers System

There are ways to make it easier—for example, the "near numbers" system. To begin, we get a *near* answer by rounding off. Thus, a grocery bill of four items might be:

$$
\begin{array}{r}
59 \\
38 \\
41 \\
+67 \\
\hline
\end{array}
$$

We know the answer must be *about* the same as $60 + 40 + 40 + 70 = 210$ ($2.10).

However, we can get an *exact* answer, almost as easily, by keeping track of the amount rounded off. That is:

$$
\begin{aligned}
59 &= 60 - 1 \\
38 &= 40 - 2 \\
41 &= 40 + 1 \\
67 &= 70 - 3
\end{aligned}
$$

*The section below reprinted courtesy of *Consumer Views*, the publication of Citibank.

Now total the numbers on the right-hand side:

$$60 - 1$$
$$40 - 2$$
$$40 + 1$$
$$70 - 3$$
$$\overline{210 - 5 = 205}$$

So the *exact* answer is $2.05.

If you do this enough times, you will soon be adding double-digit numbers directly in your head. Thus, 48 + 27 is 5 less than 80. That is 75. We call this the method of "near numbers."

Multiplying? You can use the same process. For example, 6 cans of soup at 39¢ each can be figured as 6 cans at 40¢ less 6¢; that is, $2.40 − 6¢ = $2.34. Again, with a little practice you could do this in your head.

It's in the supermarket that most of us would like to be able to do some fast mental arithmetic. Everything seems to be in different sizes and prices—a bewildering array. Which is the better buy, 24 ounces of fruit juice for 39¢, or a 32-ounce container for 53¢?

The "correct" way to solve the problem is to find the price per ounce of the smaller jar (that is, 39¢ ÷ 24 oz.) and then the price per ounce of the larger jar (53¢ ÷ 32 oz.) and then compare. But there's an easier way.

The Least Common Multiple

Look at how you multiply 24 and 32—or, in other words, the multiples of 24 and 32.

24	32
1 x 24 = 24	1 x 32 = 32
2 x 24 = 48	2 x 32 = 64
3 x 24 = 72	3 x 32 = 96
4 x 24 = 96	

Have you noticed that four 24s and three 32s are both 96? In fact, 96 is called the "least common multiple" of 24 and 32.

So to compare prices, instead of considering the price of a single ounce, look at the price of 96 ounces. Four 24-ounce bottles will cost 4 X 39¢ or $1.56. Three 32-ounce bottles will cost 3 X 53¢ or $1.59. It's cheaper to buy the *smaller*.

This system also gives you an idea of the kind of savings you could realize over a series of purchases. In some cases, of course, you might decide to spend a little more because one size or shape of package is more convenient for you.

Two Tricks: Discounts and Dollar Doubling

Here's a quick way to calculate—or check—discount prices. Just subtract the discount percentage from 100 and then multiply the answer (called "the reciprocal") and the original price.

For example, the reciprocal of a 40% discount is 60; the reciprocal of a 30% discount is 70. So to figure the cost of a $6.99 record at a 30% discount, just multiply 6.99 by 70. Your answer should be $4.89.

And here's the "secret" rule of 72. To find out roughly how many years it takes to double any amount of money at any rate of interest, just divide 72 by the "effective" rate. At 6%, for instance, you double the dollars in 12 years.

If anyone offers to double your money in any number of years, divide 72 by the years to find the rate of interest.

Checking Your Arithmetic

Any arithmetic can be checked by this technique. It won't tell you that you're right but may catch you when you're wrong.

We use the basic concept that any whole number is equivalent to the number you will obtain when you add every digit in it. We call this "reducing" a number.

For example: 78 "reduces" (7 + 8) to 15 which then reduces (1 + 5) to 6. Similarly, the number 185 reduces (1 + 8 + 5) to 14, which reduces (1 + 4) to 5. Always reduce to a single digit.

Now, suppose you have multiplied 78 X 185 and your answer is 14,430. Have you made an error? Let's see. The answer (1 + 4 + 4 + 3 + 0) = 12 which further reduces (1 + 2) to 3. If the reduced answer is 3, the reduced numbers of 78 X 185 must also be 3. If they aren't, your answer 14,430 is wrong.

What do we find? The reduced number of 78 is 6 and of 185 is 5. And 6 X 5 = 30 which reduces (3 + 0) to 3. No error.

You can check addition and subtraction, too. For example, you've added two savings accounts—$3,876 and $2,938:

$$\begin{array}{r} 3,876 \\ +2,938 \\ \hline 6,714 \end{array}$$

To check that answer: The 3876 reduces to 24 which reduces to 6. The 2938 reduces to 22 which reduces to 4. The sum should be equivalent to 6 + 4 or 10, which reduces to 1. But 6714 actually reduces (6 + 7 + 1 + 4) to 18 which reduces (1 + 8) to 9. It needs to be equivalent to 1, so the addition as shown—6714—is wrong.

Main Points

1. To easily compute your daily requirements of protein/calories/carbohydrates, read food labels for amounts of these nutrients and then plan your meals accordingly.
2. Use the listed weights and measures (metric system, too) to decide how many portions you will need for your P/C/C plan.
3. Label reading helped Paul L. prepare a tasty P/C/C plan that melted away 70 overweight pounds and restored him to youthful health.
4. Unit pricing and simple mathematics can help make you a smart, economical, and healthy shopper!
5. Study the chart on page 64 and 65 for tips on how to store food you've just bought in the most healthful manner possible.

INTO THE CUPBOARD, REFRIGERATOR . . . OR FREEZER

There's no point to menu planning, careful shopping, and ideas of proper cooking if you don't give the foods you've selected the care they require once they've entered your kitchen. Pop them into the freezer, refrigerator, or proper shelf quickly.

KEEP . . .	ON THE SHELF	IN THE REFRIGERATOR	IN THE FREEZER
Dairy Foods	(in cool, dry spot)	• keep clean, closed cartons or bottles of fresh milk in coldest part (40°F.)	• cream with 40% butterfat can be stored in liquid-tight cartons for a few months
	• tightly closed jars or packages of grated dry cheese	• keep butter covered	• soft cheeses at just the right stage of ripeness wrapped in air-tight foil. Bring to room temperature to serve
	• unopened jars of processed cheese	• tightly covered containers of cheese spreads and cottage cheese	
	• tightly closed packages of dry milk products	• wrap hard cheeses in moisture-proof coverings. Will store several months	• ice cream in original, tightly closed containers, wrapped in parchment paper or foil, will keep much longer at 0°F. than in the refrigerator
	• unopened cans of processed milks	• keep ice cream in freezing unit for a few hours to a few days; store in carton or place in clean, dry ice cube tray, cover tightly with foil. Keep temperature control at setting cold enough to keep ice cream until used	
Meat, Fish, Poultry, Eggs, Cheese, Dry Beans, and Peas	• any canned meats not labeled "to be refrigerated"	• unopened canned meats will keep for several weeks	• wrap in moisture-vapor-proof wrappings in the amounts you would cook at one time—
	• jars or cans of fish and poultry	• keep fresh meat loosely wrapped 5-8 days; ground meat 2-3 days; cooked meats 3-4 days	ground meats 1-3 months
	• dried beans and peas in air-tight wrappers in cool, dry spot	• keep eggs in refrigerator. Will store about 1 week	pork 1-4 months
	• for cheese—see above		lamb, veal 6-9 months
			beef 9-12 months

KEEP . . .	ON THE SHELF	IN THE REFRIGERATOR	IN THE FREEZER
Vegetables and Fruits	• canned foods may be kept unopened on shelf for 1 year if there are no extremes of temperature • keep glass-packed foods in air-tight containers in cool, dry spot • keep dried foods in closed wax paperboard or glass containers • store onions, potatoes, and similar foods in cool, dry area where air circulates freely	• tightly cover left-overs and store immediately . . . reuse as soon as possible • store most raw vegetables in refrigerator drawer or wrap in moisture-proof wrappings • store fresh berries in shallow, loosely covered containers. Wash just before using to avoid spoilage in storage • store frozen foods for 5 to 7 days in freezing unit	• freeze your favorite casseroles and other recipes. Most cooked foods take readily to the freezer • left-overs in air-tight containers for as long as 4 weeks • fruits and vegetables may be kept at 0°F. for at least 6 months
Breads and Cereals	• keep breads in original wrappers or in heavy waxed paper in cool, dry place; bread boxes should have air holes and be kept clean • cereals, macaroni products, and flours should be well covered in cool, dry place	• breads stale quickly at refrigerator temperatures • refrigerate baked products during hot weather only, to prevent mold	• all breads and most baked goods freeze well, keep 9-12 months • wrap in moisture-vapor-proof coverings • breads that have been frozen do not stale as quickly as fresh breads

65

4 / Fresh Foods: How to Buy, Store, Cook

Your body can handle protein, calories, and carbohydrates more efficiently in the presence of balanced nutrients from a variety of foods. When your body wants to metabolize P/C/C, it calls upon vitamins, minerals, enzymes, and other elements from eaten foods. Your body uses these nutrients to help P/C/C become better assimilated to help your body enjoy better health and youthful reactions. Fresh foods are prime sources of those nutrients needed for P/C/C assimilation into your body. Here are guidelines in helping you get the most out of fresh foods in terms of health value.

FRESH FRUITS AND VEGETABLES

These are usually best in quality and lowest in cost when in season. Whatever fruit or vegetable you buy, look first for quality. Fruits should be fresh and blemish free. They should have unbroken skins and a rich color. Vegetables should have a bright color and have as few bruised parts as possible. Fresh vegetables may need some trimming, peeling, or scraping before they are cooked or served to remove damaged leaves, bruised spots, skins, and inedible parts.

When Fresh Fruits and Vegetables Are in Season

	Jan.	Feb.	Mar.	Apr.	May	June	July	Aug.	Sept.	Oct.	Nov.	Dec.
Apples	x	x	x	x					x	x	x	x
Apricots						x	x					
Artichokes				x	x							
Asparagus				x	x	x						
Avocados	x	x	x	x								
Bananas	x	x	x	x	x	x	x	x	x	x	x	x
Beets							x					
Blueberries						x	x	x				
Broccoli	x	x	x	x	x					x	x	x
Brussels Sprouts										x	x	x
Carrots	x	x	x	x	x	x	x	x	x	x	x	x
Cauliflower	x									x	x	x
Canteloupes						x	x	x				
Celery	x	x	x	x	x	x	x	x	x	x	x	x
Cherries					x	x	x	x				
Corn						x	x	x	x			
Cranberries										x	x	x
Cucumbers						x	x	x				
Eggplant								x	x			
Grapefruit	x	x	x	x	x							
Lemons	x	x	x	x	x	x	x	x	x	x	x	x
Lettuce	x	x	x	x	x	x	x	x	x	x	x	x
Nectarines						x	x	x				
Okra							x					
Onions, Dry	x	x	x	x		x	x	x	x	x	x	x
Onions, Green					x	x	x	x	x			
Oranges	x	x	x		x	x	x	x	x	x	x	x
Peaches						x	x	x				
Pears								x	x	x	x	
Parsnips	x										x	x
Peppers							x	x	x			
Pineapples				x	x							
Plums-Prunes						x	x	x	x			
Potatoes	x	x	x	x	x	x	x	x	x	x	x	x
Radishes					x	x	x					
Rhubarb		x	x	x	x	x						
Spinach	x		x	x	x					x	x	x
Squash					x	x	x	x	x	x	x	
Strawberries				x	x							
Tangerines	x											x
Tomatoes							x	x	x			
Watermelons						x	x	x				

Sources: USDA Home & Garden Bulletin 141, 143; United Fresh Fruit & Vegetable Association.

STORAGE GUIDE FOR FRUITS AND VEGETABLES

Hold at room temperature until ripe; then refrigerate, uncovered:

Apples
Apricots
Avocados
Berries
Cherries

Grapes
Melons, except water-
   melons
Nectarines

Peaches
Pears
Plums
Tomatoes

Store in cool room or refrigerate, uncovered:

Grapefruit
Lemons

Limes
Oranges

Store in cool room, away from bright light:

Onions, mature
Potatoes

Rutabagas
Squash, winter

Sweet potatoes

Refrigerate, covered:

Asparagus
Beans, snap or wax
Beets
Broccoli
Cabbage
Carrots

Cauliflower
Celery
Corn, husked
Cucumbers
Greens
Onions, green

Parsnips
Peas, shelled
Peppers, green
Radishes
Squash, summer
Turnips

Refrigerate, uncovered:

Beans, lima, in pods
Corn, in husks

Peas, in pods
Pineapples

Watermelons

TEMPERATURE GUIDE TO FOOD SAFETY

°F

250

240 — Canning temperatures for low-acid vegetables, meat, and poultry in pressure canner.

Canning temperatures for fruits, tomatoes, and pickles in waterbath canner.

212

Cooking temperatures destroy most bacteria. Time required to kill bacteria decreases as temperature is increased.

165

Warming temperatures prevent growth but allow survival of some bacteria.

140

125 — Some bacterial growth may occur. Many bacteria survive.

Foods held more than 2 hours in this zone are subject to rapid growth of bacteria and the production of toxins by some bacteria.

60

Some growth of food poisoning bacteria may occur.

40

32 — Cold temperatures permit slow growth of some bacteria that cause spoilage.

Freezing temperatures stop growth of bacteria, but may allow bacteria to survive. (Do not store food above 10°F for more than a few weeks.)

0

FOR FOOD SAFETY
KEEP HOT FOODS HOT
COLD FOODS COLD

UNITED STATES DEPARTMENT OF AGRICULTURE • OFFICE OF COMMUNICATION

Reprinted courtesy of Parker *Natural Health Bulletin,* West Nyack, N.Y. 10994, Volume 5, No. 14. Available by subscription.

MEAT, POULTRY, AND FISH

Select good quality meat that is fresh and wholesome. Plan to use as soon after purchase as possible. The price you pay per pound of meat is not necessarily a measure of its nutritive value. The cheaper cuts and grades of lean meat can be just as full of food value as the higher-priced steaks and chops. The protein in one is just as valuable as in the other. The main difference is that cheaper cuts require greater skill in cooking and seasoning.

Meats shrink in weight and volume as they cook. Much of the total loss is water, which evaporates or goes into the drippings. Some fat also is in the drippings. The protein value of meat is not destroyed by cooking and only small amounts of it go into the drippings. Even when meats and poultry are stewed in large amounts of water, not more than 10 percent of the protein passes from meat to broth. Roasting beef to the rare state conserves more nutrition than cooking it to the well-done stage.

Fish is usually available by the pound in such forms as whole, dressed, in steaks, fillets, and chunks. Most fish dealers will clean, dress, or fillet your fish for you. Plan to eat fish often for its high-protein content. After purchase, plan to cook it as soon as possible. You may store it in your freezer for two or three days at the most, and still retain top quality.

Storage Suggestions. Store all fresh meat, poultry, and fish either in your freezer or in the coldest part of your refrigerator. Plan on a maximum of three days of storage before use.

DAIRY PRODUCTS

Fresh milk and cream. Store in refrigerator at about 40° F. Milk and cream are best stored only three to five days. Keep covered so they won't absorb odors and flavors of other foods.

Cheese spreads and cheese foods. After containers of these foods have been opened, cover and store them in your refrigerator.

Hard cheeses such as Cheddar, Parmesan, and Swiss. Keep in your refrigerator. Wrap tightly to keep out air. The original packagings should be used, when possible. Stored this way, hard cheeses will keep for several months unless mold develops.

Soft cheeses such as cottage, cream, Camembert. Store tightly covered. Use cottage cheese within three to five days, others within two weeks.

Eggs

Buy graded eggs in cartons at a store that keeps them in refrigerated cases. At home, store promptly in your refrigerator. Plan to use eggs within a week after purchase. If held too long, the thick white may thin, the yolk membrane may weaken and break when the shell is opened. Cover leftover yolks with cold water and store in your refrigerator in a covered container. Extra egg whites should also be refrigerated in a covered container. Use leftover yolks and whites within a day or two. Cooking losses in eggs are not high because the cooking period is short and the temperatures are fairly high; the protein values are well retained.

BREADS, CEREALS, GRAINS

These are prime sources of plant protein as well as calories and carbohydrates, together with other nutrients. Whole grain products retain the germ and outer layers of the grain where the nutrients are highly concentrated. You would do well to use whole grain flours, brown rice, whole ground cornmeal, whole ground oats or oatmeal,as well as whole grain bread and other products made from them.

Storing Breads and Cereals. Store in original wrapper or packaging in your breadbox or refrigerator. Breads will retain their good quality for two to three months if frozen in their original wrappers and stored in your freezer. Most cereals, flours, and dry mixes will retain quality and freshness if stored at room temperatures in tightly closed containers that keep out dust, moisture, and insects.

FATS AND OILS

Most fats and oils need protection from air, heat, and light. Fats and oils in partially filled containers keep longer if they are transferred to smaller containers in which there is little or no air space.

Butter, Margarine. Store tightly wrapped or covered in your refrigerator. Keep only as much as you need for quick use. Don't let butter or margarine stand for long periods at room temperature;

exposure to heat and light hastens rancidity. Refrigerate opened jars of salad dressing; do not freeze.

Cooking and Salad Oils. Keep small quantities at room temperature and use before flavor changes. For long storage, keep oils in your refrigerator. Some of these oils may cloud and solidify in your refrigerator. This is not harmful. If warmed to room temperature, they will become clear and liquid.

Mayonnaise and Other Salad Dressings. Keep all home-made salad dressings in your refrigerator. Purchased mayonnaise and other ready-made salad dressings should be refrigerated after jars have been opened.

WISE COOKING FOR BETTER P/C/C BALANCE

Help your body make the best use of protein, carbohydrates, and calories in a healthy balance by following these simple rules of cooking:

Plant Foods. Wherever possible, plan to eat *raw* fruits and vegetables for top quality P/C/C balance. If you must cook, then do so in as little water as possible and as quickly as possible, in a tightly covered kettle. Serve as soon after cooking as possible.

Meat Foods. When cooking meat, poultry, and fish, use baking, broiling, or roasting for maximum P/C/C balance. *Frying* and *boiling* often render nutrients insoluble; that is, these processes tend to coat protein, carbohydrates, and calories with fat, making them difficult to digest and hard to assimilate. Unless absolutely necessary, frying and boiling should *not* be used in order to avoid coating nutrients with heavy fat.

Eggs. Soft-boiled, poached, baked with other foods, eggs are a prime source of P/C/C in a healthy balance. For best P/C/C power, eggs should be cooked with low to moderate heat, for just the right amount of time. If cooking temperature is too high or the egg is cooked too long, the white becomes tough, the yolk becomes mealy, and the P/C/C ratio is upset.

Grain Foods. Whole grain breads should be baked until done, then promptly removed from oven and set out to cool. Cereals should *not* be cooked in excessive amounts of water; when you drain off cooking water, you also lose much nutrient value, along with protein, carbohydrates, and calories. Freshness of products and early use after cooking will help give you needed P/C/C balanced intake.

HOW TO PROTECT YOURSELF AGAINST
FOOD CONTAMINATION

Foods can cause illness when they contain a disease-producing agent, such as bacteria or toxin-forming mold. Bacterial contamination of foods can be reduced through either cooking or refrigeration. A rule-of-thumb is to refrigerate perishable foods as soon after purchase as possible. Here are some other suggestions:

1. Buy perishable foods—including those that are potentially hazardous because of spoilage—in small quantities so you will not keep them too long before use.

2. If you're doing other shopping, make the food market your last stop. Take foods home immediately.

3. Always put refrigerated and frozen foods in your shopping basket last.

4. At home, immediately put perishable foods in your refrigerator.

5. Make sure your refrigerator temperature is cold enough to keep foods safe.

6. Store foods in a small, shallow container. The more surface that is exposed to the cold, the faster your food will cool.

7. Keep food containers covered so that food particles from the shelf above will not fall into food stored below.

8. Never leave leftovers on your table after a meal, but store them in your refrigerator immediately.

9. Cooked foods should be served as soon as possible. Avoid prolonged keeping at room temperature. When warm-holding is necessary, the temperature should be kept at or above 140° F to prevent bacterial growth.

10. If cooked foods are to be kept for later use, they should be refrigerated *immediately*, without preliminary cooling.

WHEN YOU HANDLE FOODS . . .

1. Clean your hands before and after handling raw foods. Wash thoroughly all cutting surfaces and utensils after each use.

2. Put perishable and frozen foods in your refrigerator as soon as you get home after shopping with a minimum of handling.

3. If you prepare foods ahead of time, put them in your refrigerator until you're ready to serve them.

4. Don't leave leftover foods on the table but store them in your refrigerator promptly.

5. Keep your refrigerator at 40° F or below.

6. Don't let frozen foods thaw at room temperature. Defrost them in your refrigerator.

7. Cool foods promptly in small quantities and shallow layers so that the temperature of the food is brought down to refrigerator temperature (40° F) in two to three hours.

8. When cooking meats, use a meat thermometer to make sure the interior part of the meat is cooked thoroughly. For example: at least 175° to 185° F for poultry, 170° for fresh meat.

9. If you do home canning, use the "cold pack" method only for acid foods. Remember, when canning nonacid foods, such as meats or vegetables, you cannot be sure that temperatures used in the "cold pack" are high enough to kill the bacteria that may cause botulism (food poisoning, usually from home-canned meat and low-acid canned foods such as string beans, beets, and corn, prepared by methods other than pressure canning).

10. Be careful in the handling of pets in your home. Pet feeding dishes, toys, or bedding should not be allowed in the kitchen or near any items which come in contact with the family's food or with utensils or working surfaces used in the preparation of food.

Fresh foods are prime sources of P/C/C. When you buy, store, and cook them properly, they can do much to help keep your health in tip-top shape.

In Review

1. Nutrients in fresh foods help promote better assimilation of protein, carbohydrates, and calories.
2. Get the most from fresh fruits, vegetables, meat, poultry, fish, dairy products, eggs, breads, cereals and grains, and fats and oils, with proper care and use.
3. Follow the charts for wise buying and storage.
4. Protect yourself against food spoilage and poisoning with the simple rules and suggestions.

5 / *Frozen Foods:* How to Buy, Store, Cook

Frozen foods are more than just "something to eat." They are prime sources of many nutrients, including protein. Whether fresh from orchard, sea, or farm, frozen foods have been flash-frozen at many degrees below zero at the very peak of quality. That's why you enjoy them so much and why they can help you plan your P/C/C program. Frozen foods are rarely "out of season," and you can plan almost any type of food for your program by making proper selections from your store's freezer.

HOW TO BUY . . .

1. Before you go shopping, check how much freezer space you have available to fill with new purchases. It won't make sense to bring back more frozen goods than your freezer can hold.

2. Make your selections from your grocery store's freezer case last, *after* you've popped all your other purchases into your shopping

cart. To help yourself, follow this wise procedure—put your frozen foods needs at the end of your shopping list.

3. If your market has insulated bags available, ask the checker or bagger to be sure to put all your frozen foods in one. When you get your groceries to your car (if you are driving), place the bags with the frozens in the coolest spot, and out of the glare of the sun if possible.

4. Bring your frozen food purchases home as reasonably fast as you can.

5. As soon as you are home, put the frozen foods into your freezer *immediately*.

6. *Rotate* the frozen foods. Set up a system for yourself, whether by a check-off list you keep or simply by moving older purchases forward when you put your new purchases in. Follow this first in, first out method as much as you can. Remember, you're really buying the foods to eat and enjoy on your P/C/C program, not to store for a long time. All foods should be used within a reasonable length of time.

7. Don't take packaging lightly. To retain highest quality in frozen food, packaging materials should be moisture-vapor-proof. Glass, metal, and rigid containers are examples of this type. Most bags, wrapping materials, and waxed cartons made especially for freezing are moisture-vapor-resistant enough to retain satisfactory quality.

Note: Unacceptable packaging for freezer storage includes ordinary waxed papers, lightweight household aluminum foil, and unwaxed or lightly waxed cartons.

HOW TO STORE . . .

Scientific research (the Time-Temperature Tolerance Studies of Frozen Foods conducted by the U.S. Department of Agriculture) has shown that *zero degrees Fahrenheit ($0°F$)* is the temperature at which frozen foods should be stored after initial processing until preparing-and-serving time so that their prime condition is maintained.

Basic Rule. For peak quality in frozen foods, keep your freezer at $0°F$ or below.

FREEZER STORAGE–THE COLDER, THE BETTER

Frozen foods require low storage temperatures to preserve their quality. Foods held at 15°F may feel very hard, but are less solidly frozen than food at 0°F. Held for the same length of time, the 0°F food will be superior in color, flavor, texture, and P/C/C balance.

You can take your freezer's temperature regularly. Purchase an inexpensive refrigerator/freezer thermometer in a department store's kitchenware section and follow these simple directions:

1. Place the refrigerator/freezer thermometer in front of the storage area, fairly high up in the food load.

2. Leave the thermometer in place, overnight at least, without opening the freezer, before taking your first reading. If this reading is above 0°F, adjust the appliance's temperature control to bring it as close to a zero degree temperature as possible. Wait overnight again to take your new and colder reading.

3. Get into the habit of reading the thermometer in place daily immediately after opening the freezer.

4. Do not make your temperature check while the freezer is defrosting. If yours is a frost-free appliance, this simply means reading the thermometer when the fan is on, and not when the freezer is in its automatic defrosting cycle.

HOW TO COOK . . .

To serve, you begin by reading the instructions. Also, note these suggestions:

Thawing. It's best to thaw frozen fish, poultry, or meat in your refrigerator. That way the surface does not reach dangerously high bacteria levels before the product thaws in the center. Another benefit of slower thawing is less moisture loss (drip). Meat, fish, and poultry can be cooked without thawing; allow about one-third to one-half more cooking time.

Refreezing. Most partially thawed foods refreeze safely if they still contain ice crystals and are firm in the center. However, many foods (partially thawed ice cream) will not be top quality. Meat, fish, and poultry purposely thawed in your refrigerator and kept no more than one day may be refrozen. Do *not* refreeze thawed meat or poultry pies or casseroles, cream pies or vegetables.

Completely Thawed. When frozen foods are warmed to room temperature (72° F), they should be thoroughly cooked immediately or discarded. Fruit and fruit juice concentrates are exceptions; they ferment when spoiled so you can rely on taste to warn you. Discard them if you taste an "off" flavor.

Thawed Foods Previously Frozen. Some foods, especially meats and poultry—though purchased thawed—may have been previously frozen. If refrozen, the quality should be the same since they were commercially frozen.

Generally speaking, the faster the freezing rate, the better the P/C/C quality. Commercial freezing is much faster than home freezing. With quick freezing, there is less breakdown of cells. When water (component of all foods) freezes rapidly, tiny crystals are formed. Slower freezing forms large ice crystals which cause cells to break. Moisture leaks out and quality is lowered. So commercially frozen foods offer better P/C/C quality than home frozen foods.

SUGGESTED HOME STORAGE PERIODS TO MAINTAIN HIGH-QUALITY FROZEN FOODS STORED AT 0°F OR LOWER

Food	Months
FRESH MEATS	
Beef and lamb roasts and steaks	8 to 12
Veal and pork roasts	4 to 8
Chops, cutlets	3 to 6
Variety meats	3 to 4
Ground beef, veal or lamb, and stew meats	3 to 4
Ground pork	1 to 3
Sausage	1 to 2
CURED, SMOKED AND READY-TO-SERVE MEATS	
Ham—whole, half, or sliced	1 to 2
Bacon, corned beef, frankfurters, and wieners	Less than 1
Ready-to-eat luncheon meats	Freezing not recommended
COOKED MEAT	
Cooked meat and meat dishes	2 to 3
FRESH POULTRY	
Chicken and turkey	12

Food	*Months*
Duck and goose	6
Giblets	3

COOKED POULTRY

Cooked poultry dishes and cooked poultry slices or pieces covered with gravy or broth	6
Fried chicken	4
Sandwiches of poultry meat and cooked slices or pieces not covered with gravy or broth	1

FRESH FISH .... 6 to 9

COMMERCIALLY FROZEN FISH

Shrimp and fillets of lean type fish	3 to 4
Clams, shucked, and cooked fish	3
Fillets of fatty type fish and crab meat	2 to 3
Oysters, shucked	1

FRUITS AND VEGETABLES, most .... 8 to 12
Home-frozen citrus fruits and juices .... 4 to 6

MILK PRODUCTS

Cheddar type cheese—one pound or less, not more than one inch thick	6 or less
Butter and margarine	2
Frozen milk desserts, commercial	1

PREPARED FOODS

Cookies	6
Cakes, prebaked	4 to 9
Combination main dishes and fruit pies	3 to 6
Breads, prebaked and cake batters	3
Yeast bread dough and pie shells	1 to 2

You can enjoy a wide variety of frozen foods throughout the year. Include them in your Protein/Calorie/Carbohydrate program for better health and vigor.

Important Points

1. Keep frozen foods frozen from the time of purchase until you are ready to use them. Buy them last on your shopping list, get them home quickly, put them in your refrigerator-freezer immediately upon arrival.

2. Frozen foods will retain their peak P/C/C values if you store them properly.
3. Use a wide variety of frozen foods throughout the year to balance your P/C/C program.

6 / Canned Foods: How to Buy, Store, Cook

Canning, one of the most familiar forms of food preservation, makes an important contribution to the variety, quality, and safety of your food supply. With nutrition labeling, you can easily plan your daily amount of Protein/Calorie/Carbohydrate intake. You decide how many nutrients you require, then you make simple calculations based upon the listings, and prepare meals to fit your needs. The all-important label serves as the "window" of the can and helps you plan your P/C/C program.

HOW TO BUY . . .

The can should be firm and smooth. Avoid cans that show damage since these may be toxic. The basic buying tips include:

Bulging Cans—spoiled! DO NOT USE!
Dented Cans—do not buy cans with dents on the side seam of the can itself or on the rim seams at the top or bottom of the can.

Check carefully for leakage, especially around the seam. Throw leaky cans away.

Rusty Cans—check for leakage. The rust may have penetrated the can. If in doubt, do not use.

Read the Label for Contents. It should tell you the variety, style, and packing medium of the product to suit your taste. For instance, you want to know whether you are buying cream style or whole kernel corn. Or are you buying peaches packed in water, light syrup, or heavy syrup? The label will help you plan your meals, too. If you're making beet salad, you might want sliced beets. If you plan to make borscht, you might prefer shoestring or diced beets. Reading the label will help you tailor your shopping to your exact purpose. The label will also tell you the amount of protein, calories, and carbohydrates in the product and this helps you plan your P/C/C program, together with good taste.

If you read canned food labels over a period of time, you will learn by comparing labels which ones give you the kind of facts that you will find most helpful, and which labels are easiest to read and best arranged for you. Labels help you plan your daily P/C/C program with ease and convenience.

Canned foods have eliminated the problem of limited availability, making it possible to have a year-round supply of almost any product imaginable. This is because the foods are canned at their most bountiful time and will keep for extended periods of time.

Buying Tips. Buying goods in quantity often saves time and money. There are numerous convenient sizes determined by your needs. Small cans for single folks, medium sizes for an average family, and even larger sizes for parties and family use. When you buy canned foods, purchase the size that is best for you; it isn't a good buy, even at a lower price, if the food is not used. The adjacent chart is your guideline for planning your P/C/C program with canned foods according to the amount of food you will need.

HOW TO STORE . . .

Canned foods *do* have a long shelf life, but *don't* neglect them for several years and suddenly decide to use them. These aged canned goods may not always be safe. Not only that, but their color, flavor, texture, and/or nutritive value may have deteriorated. So plan to use them within a reasonable length of time.

CONTAINER			PRINCIPAL PRODUCTS
Industry Term	*Consumer Description* Approx. Net Weight or Fluid Measure (Check Label)	Approx. Cups	
8 oz.	8 oz.	1	Fruits, vegetables, specialties for small families. 2 servings.
Picnic	10½ to 12 oz.	1¼	Mainly condensed soups. Some fruits, vegetables, meat, fish, specialties. 2 to 3 servings.
12 oz. (vac.)	12 oz.	1½	Principally for vacuum pack corn. 3 to 4 servings.
No. 300	14 to 16 oz. (14 oz to 1 lb.)	1¾	Pork and beans, baked beans, meat products, cranberry sauce, blueberries, specialties. 3 to 4 servings.
No. 303	16 to 17 oz. (1 lb. to 1 lb., 1oz.)	2	Principal size for fruits and vegetables. Some meat products, ready-to-serve soups, specialties. 4 servings.
No. 2	20 oz. (1 lb., 4 oz.)	18 fl. oz. (1 pt., 2 fl. oz.)	2½	Juices, ready-to-serve soups, some specialties, pineapple, apple slices. No longer in popular use for most fruits and vegetables. 5 servings.
No. 2½	27 to 29 oz. (1lb., 11 oz. to 1lb., 13 oz.)	3½	Fruits, some vegetables (pumpkin, sauerkraut, spinach and other greens, tomatoes). 5 to 7 servings.
No. 3 cyl. or 46 fl. oz.	51 oz. (3 lb. 3 oz.)	46 fl. oz. (1 qt. 14 fl. oz.)	5¾	Fruit and vegetable juices, pork and beans. Institutional size for condensed soups, some vegetables. 10 to 12 servings.
No. 10	6½ lb. to 7 lb. 5 oz.	12-13	Institutional size for fruits, vegetables, and some other foods. 25 servings.

Meats, Poultry, Fish, and Seafood are almost entirely advertised and sold under weight terminology.

Store canned foods in a dry place at moderately cool, but not freezing temperatures. Rotate foods. Try not to keep canned foods over one year. Canned fruit juices should not be kept over nine months. The shelf life will be shorter when canned goods are stored above 70°F. A slight breakdown of texture may result from freezing some canned foods and this should be avoided.

How to Store Unopened Canned Foods. Store cans in your coolest cabinets, away from appliances which produce heat. Keep in a cool place for a maximum of 12 months.

Canned Fruit Juices. Store in a cool place (about 70°F) for a maximum of nine months.

How to Store Opened Canned Foods. Transfer to glass or plastic storage container. Cover tightly. Refrigerate. Fish and seafood should be used within two days. Fruit should be used within one week. Meats used within two days. Poultry within two days. Vegetables within three days. Fruits within three days.

HOW TO USE . . .

Preparation is easy with a can of food. If required, just heat and serve. Overcooking of canned or any other type of food can lead to nutrient loss. So cook only until palatable to help protect against nutrient loss and P/C/C imbalance. There are more than 1500 different foods and combinations of foods available in cans (and jars). This huge selection comes in various styles of pack—whole, sliced, julienne, diced, creamed, sauce, paste, juice, in syrup, in its own juice, in water, vacuum pack—choose the style that's right for your needs.

Many canned foods are in reality complete dishes ready to serve as they come from the can after being chilled, heated, or otherwise prepared; other canned foods may need additional seasonings or other individual touches to give them just the right taste to suit each palate. Canned foods can also provide a basis for almost endless variations and combinations of different foods to make new dishes that have appetite appeal.

When you plan your daily intake of Protein/Calories/Carbohydrates you have a variety of different foods to create endless dishes to suit endless tastes. Simply read the label. Calculate the amount of protein, calories, and carbohydrates you require and then make selections according to your needs.

When You Shop. Start out by having your daily P/C/C list with you. Then you should know, in advance, the meal you will be preparing. For example, solid pack tuna may be desirable for a salad plate, but chunk tuna would save time when making a casserole or sandwich filling. You may prefer to make your own spaghetti sauce, but at another time you would prefer to buy the prepared sauce with or without meat. Read the label to calculate your P/C/C counts and your tastes.

Canned foods can be helpful in planning P/C/C counts and should be part of your good eating program for better health.

Main Points

1. Canned foods can be useful for your Protein/Calories/Carbohydrates planning program and should form part of your planned menu.
2. Protect your health by selecting firm, smooth cans and not those that are bulging, dented, or otherwise damaged.
3. Read the label to plan your daily protein, calorie, and carbohydrate needs, along with other nutritional benefits.
4. Plan on minimal storing of canned foods to help protect against nutrient loss.
5. Use a wide variety of canned foods to suit your tastes and fulfill your daily P/C/C counts.

BRAND NAME LISTING OF PROTEIN, CALORIES, AND CARBOHYDRATES
Including Basic Foods

HOW TO USE YOUR PROTEIN, CALORIES, AND CARBOHYDRATES CHARTS

1. The first column lists the brand name or basic food according to category. Find the category and then find your specific food in alphabetical order.

2. The second column lists the quantity or measurement of the specific food.

3. The third column lists the amount of calories in the specified amount of food.

4. The fourth column lists the amount of protein in the specified amount of food.

5. The fifth column lists the amount of carbohydrates in the specified amount of food.

6. The sixth column lists the computerized ratio of the nutrients according to the most significant amount.

Finally, the computerized figures identify the following:

HP	=	HIGH PROTEIN
LP	=	LOW PROTEIN
HC	=	HIGH CALORIES
LC	=	LOW CALORIES
HCR	=	HIGH CARBOHYDRATES
LCR	=	LOW CARBOHYDRATES

A zero or 0 indicates there is an insignificant amount of the nutrient present, if at all, and is too small to be considered of importance.

A dash (−) indicates the amount is either unavailable or undetermined and is to be considered of small importance.

Alcoholic Beverages

Food	Quantity	Calories	Protein Grams	Carbo-hydrates Grams	P/C/C Com-puter
Beer, alcohol 4.5% by volume	3½ oz.	42	.3	3.8	HC
Gin, 80 proof	3½ oz.	231	–	Trace	HC
Malt, dry	3½ oz.	368	13.1	77.4	HC
Malt extract, dried	3½ oz.	367	6.0	89.2	HC
Rum, 80 proof	3½ oz.	231	–	Trace	LCR
Vodka, 80 proof	3½oz.	231	–	Trace	LCR
Whiskey, 80 proof	3½ oz.	231	–	Trace	LCR
Wine, dessert, 18.8% alcohol	3½ oz.	137	.1	7.7	LCR
Wine, table, 12.2% alcohol	3½ oz.	85	.1	4.2	LCR

Beverages

Food	Quantity	Calories	Protein Grams	Carbo-hydrates Grams	P/C/C Com-puter
Acerola juice	3½ oz.	23	.4	4.8	LC
Apple juice, canned or bottled	3½ oz.	47	.1	11.9	LC
Birds Eye Frozen Awake Juice	8 oz.	110	0	25.2	HCR
Birds Eye Frozen Orange Juice	8 oz.	102	1.8	26.6	HC
Birds Eye Orange Plus	8 oz.	134	1.2	33.2	HCR
Blackberry juice, canned, unsweetened	3½ oz.	37	.3	7.8	LCR
Bouillon cubes or powder	3½ oz.	120	20	5	HP
Campbell's Bean w/Bacon Soup	10 oz.	200	8	26	HC
Campbell's Beef Broth	10 oz.	35	4	3	LC
Campbell's Beef Consommé Broth	10 oz.	40	5	3	LC
Campbell's Beef Noodle Soup	10 oz.	90	5	10	HC
Campbell's Beef Soup	10 oz.	100	7	14	HC
Campbell's Bisque Tomato Soup	10 oz.	150	2	27	HC
Campbell's Black Bean Soup	10 oz.	130	6	22	HC
Campbell's Cheddar Cheese Soup	10 oz.	180	6	12	HC
Campbell's Chicken Broth	10 oz.	50	6	2	LC
Campbell's Chicken 'n Dumplings Soup	10 oz.	120	6	7	HC
Campbell's Chicken Gumbo Soup	10 oz.	70	2	10	LC
Campbell's Chicken Noodle Soup	10 oz.	90	4	11	LCR
Campbell's Chicken Noodle-O's Soup	10 oz.	90	4	12	LCR
Campbell's Chicken w/Rice Chunky Soup	9½ oz.	160	14	16	HC
Campbell's Chicken w/Rice Soup	10 oz.	80	4	9	LC
Campbell's Chicken and Stars Soup	10 oz.	80	4	9	LC
Campbell's Chicken Vegetable Soup	10 oz.	90	4	10	LC
Campbell's Chunky Beef Soup	9½ oz.	210	15	21	HC
Campbell's Chunky Chicken Soup	9½ oz.	200	14	20	HC
Campbell's Chunky Sirloin Burger Soup	9½ oz.	210	12	21	HC
Campbell's Chunky Split Pea Soup w/Ham	9½ oz.	220	12	30	HC
Campbell's Chunky Turkey Soup	9½ oz.	160	11	17	HC
Campbell's Chunky Vegetable Soup	9½ oz.	140	3	22	HC
Campbell's Cream of Asparagus Soup	10 oz.	100	2	12	LC
Campbell's Cream of Celery Soup	10 oz.	110	2	10	LC
Campbell's Cream of Chicken Soup	10 oz.	140	4	10	HC
Campbell's Cream of Mushroom Soup	10 oz.	150	2	11	HC

Food	Quantity	Calories	Protein Grams	Carbo-hydrates Grams	P/C/C Computer
Campbell's Cream of Potato Soup	10 oz.	90	2	14	LC
Campbell's Cream of Potato Soup w/Milk	10 oz.	140	5	18	HC
Campbell's Cream of Shrimp Soup	10 oz.	120	3	10	HC
Campbell's Cream of Shrimp Soup w/Milk	10 oz.	210	18	17	HC
Campbell's Chili Beef Soup	10 oz.	190	7	24	HC
Campbell's Curley Noodle w/Chicken Soup	10 oz.	100	4	12	LC
Campbell's Golden Mushroom Soup	10 oz.	100	3	11	LC
Campbell's Golden Vegetable Noodle-O's	10 oz.	90	2	13	LC
Campbell's Green Pea Soup	10 oz.	180	9	29	HC
Campbell's Hot Dog Bean Soup	10 oz.	210	10	25	HC
Campbell's Low Sodium Cream of Mushroom Soup	7¼ oz.	130	2	10	HC
Campbell's Low Sodium Green Pea Soup	7½ oz.	140	7	22	HC
Campbell's Low Sodium Tomato Soup	7¼ oz.	100	2	18	HP
Campbell's Low Sodium Turkey Noodle Soup	7¼ oz.	60	3	7	LP
Campbell's Low Sodium Vegetable Beef Soup	7¼ oz.	80	5	8	LP
Campbell's Low Sodium Vegetable Soup	7¼ oz.	80	2	14	LP
Campbell's Manhattan Chunky Clam Chowder	9½ oz.	160	8	23	HP
Campbell's Manhattan Clam Chowder Soup	10 oz.	100	2	15	LCR
Campbell's Minestrone Soup	10 oz.	110	5	15	LCR
Campbell's New England Clam Chowder	10 oz.	100	6	13	LCR
Campbell's New England Clam Chowder w/Milk	10 oz.	190	11	20	HP
Campbell's Noodles and Ground Beef Soup	10 oz.	110	5	14	LCR
Campbell's Old-Fashioned Tomato Rice Soup	10 oz.	140	2	25	LP
Campbell's Old-Fashioned Vegetable Soup	10 oz.	90	3	11	LP
Campbell's Onion Soup	10 oz.	80	4	10	LC
Campbell's Oyster Stew Soup	10 oz.	80	2	5	LC
Campbell's Oyster Stew Soup w/Milk	10 oz.	170	7	12	HC
Campbell's Pepper Pot Soup	10 oz.	130	7	12	HC

Food	Quantity	Calories	Protein Grams	Carbo-hydrates Grams	P/C/C Computer
Campbell's Scotch Broth	10 oz.	100	5	11	LC
Campbell's Split Pea Soup w/Ham	10 oz.	210	10	30	HCR
Campbell's Stockpot Soup	10 oz.	120	5	13	LP
Campbell's Tomato Soup	10 oz.	110	2	20	LP
Campbell's Tomato Soup w/Milk	10 oz.	210	7	27	HC
Campbell's Tomato-Beef Noodle-O's Soup	10 oz.	160	5	24	HC
Campbell's Turkey Noodle Soup	10 oz.	90	4	10	LC
Campbell's Turkey Vegetable Soup	10 oz.	90	3	10	LC
Campbell's V-8 Drink	8 oz.	47	1.3	9.3	LC
Campbell's Vegetable Beef Soup	10 oz.	90	6	10	LC
Campbell's Vegetable Soup	10 oz.	100	3	17	LC
Campbell's Vegetarian Vegetable Soup	10 oz.	90	2	16	LC
Carnation Chocolate Instant Breakfast Drink	1 pkg. + 8 oz. milk	290	16	35	HC
Carnation Coffee Instant Breakfast Drink	1 pkg. + 8 oz. milk	290	16	36	HC
Carnation Vanilla Instant Breakfast Drink	1 pkg. + 8 oz. milk	290	16	36	HC
Chocolate (hot, beverage)	3½ oz.	95	3.3	10.4	HC
Chocolate drink with skim milk	3½ oz.	76	3.3	10.9	LP
Chocolate drink with whole (3.5%) fat milk	3½ oz.	85	3.4	11.0	HC
Club soda	3½ oz.	0	0	0	LC
Cocoa (hot, beverage)	3½ oz.	97	3.8	10.9	HC
Cocoa beverage powder with non-fat dry milk	3½ oz.	359	18.6	70.8	HC
Cocoa beverage powder without milk	3½ oz.	347	4.0	89.4	HC
Cocoa, dry powder, for breakfast	3½ oz.	265	17.3	51.5	HC
Cocoa mix for hot chocolate	3½ oz.	392	9.4	73.9	HC
Coconut milk	3½ oz.	252	3.2	5.2	HC
Coconut water	3½ oz.	22	.3	4.7	LCR
Coffee, instant, beverage	3½ oz.	1	Trace	Trace	LCR
Cola type soda	3½ oz.	39	0	10	LC
Cragmont Cherry Drink	8 oz.	120	0	29	HC
Cragmont Citrus Cooler	8 oz.	120	0	29	HC
Cragmont Diet Drinks	8 oz.	0	0	0	LC
Cragmont Grape Drink	8 oz.	130	0	32	HCR
Cragmont Orange Drink	8 oz.	130	0	32	HCR
Cragmont Pineapple-Orange Drink	8 oz.	130	0	32	HCR
Cragmont Tropical Punch	8 oz.	130	0	32	HCR
Cragmont Wild Berry Drink	8 oz.	120	0	31	HCR

Food	Quantity	Calories	Protein Grams	Carbo-hydrates Grams	P/C/C Computer
Cream soda	3½ oz.	43	0	11	LCR
Del Monte Apple Drink	8 oz.	112	0.03	31.4	HCR
Del Monte Apricot Nectar	8 oz.	135	0.9	35.9	HCR
Del Monte Fruit Punch	8 oz.	110	0.2	30.1	HCR
Del Monte Grape Drink	8 oz.	120	0.02	32.8	HCR
Del Monte Grapefruit Juice, Sweetened	8 oz.	96	1.5	25.4	HC
Del Monte Grapefruit Juice, Unsweetened	8 oz.	93	1.7	23.5	HCR
Del Monte Orange Drink	8 oz.	103	0.02	28.7	HCR
Del Monte Orange Juice, Sweetened	8 oz.	115	1.5	29.9	HCR
Del Monte Orange Juice, Unsweetened	8 oz.	93	1.7	23.5	HCR
Del Monte Orange-Apricot Juice	8 oz.	118	0.5	32.3	HCR
Del Monte Orange and Grapefruit Juice, Sweetened	8 oz.	108	1.7	27.9	HCR
Del Monte Peach Nectar	8 oz.	140	1	37.9	HCR
Del Monte Pear Nectar	8 oz.	140	0.5	38.4	HCR
Del Monte Pineapple-Apricot Drink	8 oz.	132	0.07	32.6	HCR
Del Monte Pineapple-Cherry Drink	8 oz.	103	1.7	24.5	HCR
Del Monte Pineapple-Grapefruit Drink	8 oz.	125	0.05	31.4	HCR
Del Monte Pineapple-Orange Drink	8 oz.	125	0.05	31.6	HCR
Del Monte Pineapple-Pear Drink	8 oz.	135	0.05	34	HCR
Del Monte Pineapple Juice, Unsweetened	8 oz.	120	1	32.3	HCR
Del Monte Pink Pineapple-Grape-fruit Drink	8 oz.	127	0.02	32.3	HCR
Del Monte Prune Juice	8 oz.	117	1.5	31.1	HCR
Del Monte Tomato Juice	8 oz.	44	1.9	9.8	LCR
Dole Pineapple-Grapefruit Juice	8 oz.	121	0.45	31	HC
Dole Pineapple Juice	8 oz.	124	0.91	30.4	HC
Fruit-flavored soda (10-13% sugar)	3½ oz.	46	0	12	LC
Funny Face Choo-Choo Cherry Drink	8 oz.	90	0	22	LC
Funny Face Goofy Grape Drink	8 oz.	90	0	22	LC
Funny Face Jolly Olly Orange Drink	8 oz.	90	0	22	LC
Funny Face Loud Mouth Punch	8 oz.	80	0	19	LC
Funny Face Rah Rah Root Beer	8 oz.	90	0	22	HCR

Food	Quantity	Calories	Protein Grams	Carbo-hydrates Grams	P/C/C Computer
Funny Face Rootin' Tootin' Raspberry Drink	8 oz.	80	0	19	HC
Funny Face Rudi Tutti-Frutti Drink	8 oz.	80	0	20	HC
Funny Face Strawberry Freckle Face Drink	8 oz.	80	0	19	HC
Funny Face Tart Lil' Lemonade	8 oz.	80	0	21	HC
Funny Face With-It Watermelon Drink	8 oz.	80	0	20	HC
Gatorade Thirst Quencher	8 oz.	74	0	20	HC
Ginger ale	3½ oz.	31	0	8	LC
Grape juice, canned or bottled	3½ oz.	66	.2	16.6	LC
Grape juice, frozen concentrate	3½ oz.	183	.6	46.3	HCR
Grapefruit juice, canned	3½ oz.	41	.5	9.8	LC
Grapefruit juice, fresh	3½ oz.	39	.5	9.2	LC
Grapefruit juice, frozen concentrate	3½ oz.	145	1.9	34.6	HC
Grapefruit-orange juice blend, canned	3½ oz.	43	.6	10.1	LC
Heinz Great American Bean w/Ham Soup	10 oz.	231	11.97	28.2	HC
Heinz Great American Beef Noodle w/Dumplings Soup	10 oz.	125	6.8	14.5	HC
Heinz Great American Beef Stock w/Vegetable Soup	10 oz.	165	9.7	21.1	HC
Heinz Great American Chicken Gumbo Soup	10 oz.	103	3.9	15.96	LCR
Heinz Great American Chicken Noodle Soup w/Dumplings	10 oz.	105	6.3	10.5	LC
Heinz Great American Chicken Rice w/Mushrooms Soup	10 oz.	103	5.1	145.35	HCR
Heinz Great American Chili Beef Soup	10 oz.	205	9.10	25.4	HC
Heinz Great American Cream of Chicken Soup	10 oz.	128	6.5	10.5	LCR
Heinz Great American Cream of Mushroom Soup	10 oz.	151	5.1	14.25	HC
Heinz Great American Ground Beef w/Vegetable Soup	10 oz.	160	12.25	14	HC
Heinz Great American Manhattan Clam Chowder Soup	10 oz.	117	6.3	16.5	LC
Heinz Great American Split Pea Soup w/Ham	10 oz.	208	12.25	26.8	HC
Heinz Great American Tomato Soup	10 oz.	194	3.4	0.85	HC

Food	Quantity	Calories	Protein Grams	Carbo-hydrates Grams	P/C/C Com-puter
Heinz Great American Turkey Noodle Soup	10 oz.	111	6.3	14	HC
Heinz Great American Turkey Rice Soup	10 oz.	114	6	14.8	HC
Heinz Great American Turkey Vegetable Soup	10 oz.	114	6.8	14	HC
Heinz Great American Vegetable Beef Soup	10 oz.	134	11.1	14	HC
Heinz Great American Vegetarian Vegetable Soup	10 oz.	117	4.3	17.95	LP
Hi-C Apple Drink	8 oz.	116	0.09	29.5	LP
Hi-C Cherry Drink	8 oz.	123	0.03	30.26	HCR
Hi-C Citrus Cooler	8 oz.	123	0	30.26	HCR
Hi-C Grape Drink	8 oz.	117	0.03	29.1	HC
Hi-C Lemon-Lime Drink	8 oz.	109	–	27.59	HC
Hi-C Orange Drink	8 oz.	119	0.07	29.6	HC
Hi-C Orange-Pineapple Drink	8 oz.	117	0.12	29.1	HC
Hi-C Pineapple-Grapefruit Drink	8 oz.	125	0.15	31.5	HCR
Hi-C Pineapple Nectar/Alegre Mango	8 oz.	183	0.12	45.57	HC
Hi-C Punch Drink	8 oz.	131	0.05	32.13	HC
Hi-C Strawberry Drink	8 oz.	116	–	31.2	HC
Hi-C Wild Berry Drink	8 oz.	123	–	30.26	HC
Highway Tomato Juice	8 oz.	46.55	1.33	10.6	LC
Hunt's Tomato Juice	8 oz.	46	2.2	10.4	LC
Kool-Aid, Regular, All Flavors	8 oz.	98	–	25	HC
Kool-Aid, Sweetened, All Flavors	8 oz.	91	–	23	HC
Lemon juice: Canned or bottled	3½ oz.	23	.4	7.6	LC
Fresh	3½ oz.	25	.5	8.0	LC
Frozen	3½ oz.	22	.4	7.2	LC
Frozen concentrate	3½ oz.	116	2.3	37.4	HCR
Lemonade concentrate, frozen	3½ oz.	195	.2	51.1	HCR
Libby's Tomato Juice	8 oz.	45	2	10	LC
Lime juice, canned or bottled	3½ oz.	26	.3	9.0	LCR
Lime juice, fresh	3½ oz.	26	.3	9.0	LCR
Limeade concentrate, frozen	3½ oz.	187	.2	49.5	HC
Lincoln Cranberry Juice	8 oz.	144	0.2	35	HC
Malted milk beverage	3½ oz.	104	4.7	11.7	LP
Maxim Freeze-Dried Coffee	6 oz.	4	0	0.9	LC
Maxwell House Ground Coffee	6 oz.	2	0	0.4	LC
Maxwell House Instant Coffee	6 oz.	4	0	0.9	LC
Maxwell House Max-Pax Coffee Ring	6 oz.	2	0	0.5	LC
Minute Maid Snow Crop Frozen Grape Juice	8 oz.	132	0.2	33.3	HC

Food	Quantity	Calories	Protein Grams	Carbo-hydrates Grams	P/C/C Computer
Minute Maid Snow Crop Frozen Grapefruit Juice	8 oz.	100	0.56	24.4	HCR
Minute Maid Snow Crop Frozen Lemon Juice	8 oz.	53	0.39	17.6	LC
Minute Maid Snow Crop Frozen Lemon-Limeade Juice	8 oz.	100	0.06	26.1	HCR
Minute Maid Snow Crop Frozen Lemonade Juice	8 oz.	99	0.05	26.1	HCR
Minute Maid Snow Crop Frozen Limeade Juice	8 oz.	100	0.08	26.8	HCR
Minute Maid Snow Crop Frozen Orange Juice	8 oz.	120	0.76	28.5	HC
Minute Maid Snow Crop Frozen Orange-Grapefruit Blend Juice	8 oz.	101	0.66	25.5	LC
Minute Maid Snow Crop Frozen Orangeade	8 oz.	125	0.12	30.3	HC
Minute Maid Snow Crop Frozen Sweetened Tangerine Juice	8 oz.	113	0.05	27.7	HC
Minute Maid Snow Crop Sweetened Frozen Grapefruit Juice	8 oz.	113	0.51	29.7	LC
Orange juice, all varieties, fresh	3½ oz.	45	.7	10.4	LC
Canned juice	3½ oz.	48	.8	11.2	LC
Canned concentrate	3½ oz.	223	4.1	50.7	HCR
Dehydrated (crystals)	3½ oz.	380	5.0	88.9	HCR
Frozen concentrate	3½ oz.	158	2.3	38.0	HCR
Orange-apricot juice drink	3½ oz.	50	.3	12.7	LC
Pantry Pride Bean w/Bacon Soup	10 oz.	190	8	25	HC
Pantry Pride Cream of Mushroom Soup	10 oz.	160	2	14	HC
Pantry Pride Pineapple Juice	8 oz.	133	0	33.25	HC
Pantry Pride Tomato Juice	8 oz.	45	2	10	LC
Pantry Pride Tomato Soup	10 oz.	100	2	23	LC
Peach nectar, canned (approx. 40% fruit)	3½ oz.	48	.2	Trace	LC
Pear nectar, canned (approx. 40% fruit)	3½ oz.	52	.3	13.2	LC
Pillsbury Chocolate Instant Breakfast Drink	1 pkg. + 8 oz. milk	290	14	38	HC
Pillsbury Chocolate Malt Instant Breakfast Drink	1 pkg. + 8 oz. milk	280	14	37	HC
Pillsbury Strawberry Instant Breakfast Drink	1 pkg. + 8 oz. milk	280	13	38	HC
Pillsbury Vanilla Instant Breakfast Drink	1 pkg. + 8 oz. milk	280	13	39	HC

Food	Quantity	Calories	Protein Grams	Carbo- hydrates Grams	P/C/C Com- puter
Pineapple juice:					
Canned, unsweetened	3½ oz.	55	.4	13.5	LC
Frozen concentrate, unsweetened	3½ oz.	179	1.3	44.3	HC
Pineapple-grapefruit juice drink (about 40% juices)	3½ oz.	54	.2	13.6	LCR
Pineapple-orange juice drink (about 40% juices)	3½ oz.	54	.2	13.5	LCR
Postum	8 oz.	16	0.3	3.7	LC
Prune juice, canned or bottled	3½ oz.	77	.4	19.0	LC
Quinine soda, sweetened	3½ oz.	31	–	8	LC
R & R Chicken Broth	10 oz.	40	4.6	2	LC
R & R Chicken w/Rice Broth	10 oz.	60	5.1	6.3	LC
ReaLemon Reconstituted Lemon Juice	2 tbsp.	6	0	2	LC
ReaLemon Reconstituted Lime Juice	2 tbsp.	4	0	1	LC
Root beer	3½ oz.	41	0	10.5	LC
S & W Nutradiet Blue-Green Label Apricot-Pineapple Nectar	8 oz.	71	0.96	16.63	LCR
S & W Nutradiet Blue-Green Label Grape Drink	8 oz.	137	0.87	35.14	HC
S & W Nutradiet Blue-Green Label Pineapple Drink	8 oz.	135	0.71	32.35	HC
S & W Nutradiet Blue-Green Label Tomato Juice	8 oz.	50	2.06	9.55	LC
S & W Nutradiet Blue-Green Label Vegetable Juice	8 oz.	48	1.87	9.57	LC
S & W Nutradiet Red Label Apricot Nectar	8 oz.	23	1.10	16.04	LC
S & W Nutradiet Red Label Pear Nectar	8 oz.	21	0.96	15.63	LC
Sanka Freeze-Dried Coffee	6 oz.	4	0	0.8	LC
Sanka Ground Coffee	6 oz.	2	0	0.4	LC
Sanka Instant Coffee	6 oz.	4	0	0.9	LC
Sauerkraut juice, canned	3½ oz.	10	.7	2.3	LC
Slender All-Flavors Diet Drink	8 oz.	180	12.96	22.96	HC
Slender Chocolate Dry Mix Envelope	+ 6 oz. milk	225	16.25	24.3	HC
Slender French Vanilla Dry Mix Envelope	+ 6 oz. milk	225	16.25	25.2	HC
Soups, commercial, canned:	(all 3½ oz.)				
Asparagus, cream of		54	2.0	8.4	LC
Bean with pork		134	6.4	17.3	LCR
Beef broth, bouillon and consommé		26	4.2	2.2	LC

Food	Quantity	Calories	Protein Grams	Carbo-hydrates Grams	P/C/C Computer
Soups *(cont.):*					
Beef noodle		57	3.2	5.8	LC
Celery, cream of		72	1.4	7.4	LC
Chicken consommé		18	2.8	1.5	LC
Chicken, cream of		79	2.4	6.7	LC
Chicken gumbo		46	2.6	6.1	LC
Chicken noodle		53	2.8	6.6	LC
Chicken with rice		39	2.6	4.7	LC
Chicken vegetable		62	3.4	7.7	LCR
Clam chowder, Manhattan type		66	1.8	10.0	LC
Minestrone		87	4.0	11.6	LCR
Mushroom, cream of		111	1.9	8.4	HC
Onion		54	4.4	4.3	LC
Pea, green		106	4.6	18.4	HC
Pea, split		118	7.0	17.0	HC
Tomato		72	1.6	12.7	LC
Turkey noodle		65	3.6	7.0	LC
Vegetable beef		65	4.2	7.9	LC
Vegetable with beef broth		64	2.2	11.0	LC
Vegetarian vegetable		64	1.8	10.6	LC
Dehydrated:					
Beef noodle mix, dry form		28	1.0	4.8	LC
Chicken noodle mix, dry form		22	.8	3.2	LC
Chicken rice mix, dry form		20	.5	3.5	LC
Onion mix, dry form		15	.6	2.3	LC
Pea, green, mix, dry form		50	3.1	8.4	LC
Tomato vegetable with noodles mix, dry form		27	.6	5.1	LC
Frozen:					
Clam chowder, New England type		54	1.8	4.4	LC
Pea, green, with ham		57	3.8	8.0	LC
Potato, cream of		44	1.4	4.9	LC
Shrimp, cream of		66	2.0	3.5	LC
Vegetable with beef		35	2.7	3.4	LC
Start Instant Breakfast Drink	8 oz.	120	—	29.8	HCR
Stokely-Van Camp Grapefruit Juice, Sweetened	8 oz.	134	1.2	32.4	HCR
Stokely-Van Camp Grapefruit Juice, Unsweetened	8 oz.	104	1.2	24.8	HCR

Beverages

Food	Quantity	Calories	Protein Grams	Carbo-hydrates Grams	P/C/C Computer
Stokely-Van Camp Orange Juice, Sweetened	8 oz.	130	1.6	30.6	HCR
Stokely-Van Camp Orange Juice, Unsweetened	8 oz.	120	2	28.2	HCR
Stokely-Van Camp Orange and Grapefruit Juice, Sweetened	8 oz.	126	1.2	30.6	HCR
Stokely-Van Camp Pineapple Juice	8 oz.	126	1	34	HC
Stokely-Van Camp Tomato Juice	8 oz.	56	2	10.4	LC
Swanson Beef Broth	10 oz.	25	2.5	1.25	LC
Swanson Chicken Broth	10 oz.	37.5	3.75	1.25	LC
Tang Grape Drink	8 oz.	122	–	31.6	HC
Tang Grapefruit Drink	8 oz.	122	–	30	HC
Tang Orange Drink	8 oz.	122	–	29.4	HC
Tangelo juice, fresh	3½ oz.	41	.5	9.7	LC
Tangerine juice, fresh	3½ oz.	43	.5	10.1	LC
Canned	3½ oz.	43	.5	10.2	LC
Frozen concentrate	3½ oz.	162	1.7	38.3	HC
Tea, beverage	3½ oz.	2	–	4	LC
Tomato juice, canned or bottled	3½ oz.	19	.9	4.3	LC
Tomato juice, canned concentrate	3½ oz.	20	.9	4.5	LC
Tomato juice cocktail, canned or bottled	3½ oz.	21	.7	5.0	LC
Town House Apricot Nectar	8 oz.	146	1.33	37.2	HC
Town House Bean w/Bacon Soup	10 oz.	190	8	25	HC
Town House Chicken Noodle Soup	10 oz.	80	4	11	LC
Town House Chicken "O" Noodles Soup	10 oz.	90	4	11	LC
Town House Chicken w/Rice Soup	10 oz.	70	1	8	LC
Town House Chicken Star w/Noodles Soup	10 oz.	70	4	9	LC
Town House Cream of Celery Soup	10 oz.	140	2	15	HC
Town House Cream of Chicken Soup	10 oz.	120	4	12	HC
Town House Cream of Mushroom Soup	10 oz.	160	2	14	HC
Town House Pineapple Juice	8 oz.	133	0	33.25	HCR
Town House Split Pea Soup w/Ham	10 oz.	200	11	36	HC
Town House Tomato Juice	8 oz.	46.55	1.33	10.6	LC
Town House Tomato Soup	10 oz.	100	2	23	HC
Town House Turkey Noodle Soup	10 oz.	90	4	11	LC
Town House Vegetable Beef Soup	10 oz.	80	5	13	LC
Town House Vegetable Juice Cocktail	8 oz.	46.55	1.33	10.6	LC
Town House Vegetable Soup	10 oz.	80	4	13	LC

Food	Quantity	Calories	Protein Grams	Carbo- hydrates Grams	P/C/C Com- puter
Vegetable juice cocktail, canned	3½ oz.	17	.9	3.6	LC
Wyler Barley Beef-Flavored Soup Mix	8 oz.	72	2.23	12.37	LC
Wyler Beef Bouillon	6 oz.	7	11.3	11	LC
Wyler Chicken Bouillon	6 oz.	7	9.7	18.5	LC
Wyler Imitation Drink Mixes	8 oz.	90	0	22	HCR
Wyler Leek w/Potato Soup Mix	8 oz.	86	1.76	11.34	LC
Wyler Low Sodium Beef Bouillon	6 oz.	11	0.4	1.7	LC
Wyler Low Sodium Chicken Bouillon	6 oz.	11	0.31	1.6	LC
Wyler Noodle Beef-Flavored Soup Mix	8 oz.	49	1.81	9.21	LC
Wyler Noodle Chicken-Flavored Soup Mix	8 oz.	44	2.14	5.70	LC
Wyler Onion Bouillon	6 oz.	10	15.5	31	HP
Wyler Onion Soup Mix	8 oz.	38	1.05	6.91	LC
Wyler Rice Chicken-Flavored Soup Mix	8 oz.	50	1.08	8.68	LC
Wyler Vegetable Bouillon	6 oz.	7	12.7	8	HP
Wyler Vegetable Soup Mix	8 oz.	56	2.07	9.56	LC
Yuban Ground Coffee	6 oz.	2	0	0.4	LC
Yuban Instant Coffee	6 oz.	4	0	0.9	LC

Breakfast Foods, Breads, Cereals, Flours, Grains

Food	Quantity	Calories	Protein Grams	Carbo-hydrates Grams	P/C/C Computer
Almond meal, partially defatted	3½ oz.	408	39.5	28.9	HC
Argo Corn Starch	1 tbsp.	35	–	8	LC
Aunt Jemima Buckwheat Pancake and Waffle Mix	3 portions	183	6.8	22.5	HCR
Aunt Jemima Buttermilk Pancake and Waffle Mix	3 portions	251	8.6	33.7	HCR
Aunt Jemima Complete Pancake Mix	3 portions	181	4.7	35	HCR
Aunt Jemima Corn Bread Mix	1 piece	116	2.5	17.5	HC
Aunt Jemima Deluxe Easy Pour Pancake and Waffle Mix	3 portions	235	7.9	33	HC
Aunt Jemima Frozen Buttermilk Waffles	2 sections	113	2.3	15.7	HCR
Aunt Jemima Frozen French Toast	2 slices	175	6.5	27.2	HCR
Aunt Jemima Original Frozen Waffles	2 sections	114	2.3	15.7	HCR
Aunt Jemima Original Pancake and Waffle Mix	3 portions	181	6	23.5	HCR
Aunt Jemima Self-Rising Flour	¼ cup	96	2.4	20.8	HCR
Barley, pearled, light	3½ oz.	349	8.2	78.8	HC
Barley, pearled, pot or Scotch	3½ oz.	348	9.6	77.2	HC
Bean flour, lima	3½ oz.	343	21.5	63.0	HC
Bel-Air Frozen Buttermilk Waffles	1 section	100	2	19.5	LC
Bel-Air Frozen French Toast	2 slices	230	6	40	HC
Bel-Air Frozen Waffles	1 section	55	1	10.5	LC
Best Enriched Flour	¼ cup	100	2.6	21.8	LC
Biscuit dough, chilled in cans	3½ oz.	277	7.3	46.4	HC
Biscuit dough, frozen	3½ oz.	327	5.7	48.9	HC
Biscuit mix, dry form	3½ oz.	424	7.7	68.7	HC
Biscuits baked with milk	3½ oz.	325	7.1	52.3	HC
Biscuits, baking powder	3½ oz.	369	7.4	45.8	HC
Boston brown bread	3½ oz.	211	5.5	45.6	HCR
Bran with sugar and defatted wheat germ	3½ oz.	238	10.8	78.8	HC
Bran with sugar and malt extract	3½ oz.	240	12.6	74.3	HC
Bran flakes (40% bran) added thiamine	3½ oz.	303	10.2	80.6	HC
Bran flakes with raisins, added thiamine	3½ oz.	287	8.3	79.3	HC
Bread stuffing mix	3½ oz.	371	12.9	72.4	HC
Breadcrumbs, dry, grated	3½ oz.	392	12.6	73.4	HC
Brewer's yeast, debittered	3½ oz.	283	38.8	38.4	HC
Brown-and-serve rolls	3½ oz.	311	8.5	56.0	HC
Buckwheat, whole grain	3½ oz.	335	11.7	72.9	HCR
Burger Bonus Meat Extender	1/3 oz.	30.5	5.12	2.97	LCR

Food	Quantity	Calories	Protein Grams	Carbo-hydrates Grams	P/C/C Computer
Carob flour (St. John's bread)	3½ oz.	180	4.5	80.7	HCR
Chestnut flour	3½ oz.	362	6.1	76.2	HCR
Comet Brown Rice	½ cup	105	2	24	HC
Comet Long Grain Rice	½ cup	95	2	22	LC
Comet White Enriched Rice	½ cup	110	2	24	LC
Cooky dough, plain, chilled in roll	3½ oz.	496	3.9	64.9	HC
Cooky (brownie) mix	3½ oz.	403	4.9	59.8	HC
Corn bread from mix	3½ oz.	233	6.1	32.9	HCR
Corn bread with whole ground cornmeal	3½ oz.	207	7.4	29.1	HCR
Corn flakes, sugar-covered	3½ oz.	386	4.4	91.3	HC
Corn flakes, with protein concentrate	3½ oz.	378	23.0	67.0	HC
Corn flour	3½ oz.	368	7.8	76.8	HC
Corn fritters	3½ oz.	377	7.8	39.7	HC
Corn grits, enriched	3½ oz.	51	1.2	11.0	LC
Corn grits, unenriched	3½ oz.	51	1.2	11.0	LC
Cornmeal, white or yellow	3½ oz.	355	9.2	73.7	HCR
Corn pone	3½ oz.	204	4.5	36.2	HC
Corn, puffed	3½ oz.	379	4.0	89.8	HCR
Corn-rice-wheat flake cereal	3½ oz.	389	7.4	86.1	HCR
Corn, shredded	3½ oz.	389	7.0	86.9	HCR
Cornstarch	3½ oz.	362	.3	87.6	HCR
Cottonseed flour	3½ oz.	356	48.1	33.0	HC
Cracked wheat bread	3½ oz.	263	8.7	52.1	HC
Dromedary Banana Nut Sweet Bread	3½ oz.	264	6.5	43.5	HC
Dromedary Chocolate Nut Sweet Bread	3½ oz.	304	5.7	51	HC
Dromedary Corn Bread Mix	3½ oz.	323	6.5	47.8	HC
Dromedary Corn Muffin Bread Mix	3½ oz.	365	6.2	51.9	HC
Dromedary Date and Nut Sweet Bread	3½ oz.	261	5.4	42.9	HC
Dromedary Orange Nut Sweet Bread	3½ oz.	275	5.3	47.3	HC
Eggo Frozen Blueberry Waffles	1 section	130	2	18	HC
Eggo Frozen French Toast	2 slices	200	8	30	HC
Eggo Frozen Waffles	1 section	136	3.1	19.3	HC
Farina, enriched, regular	3½ oz.	42	1.3	8.7	LC
Farina, instant-cooking	3½ oz.	55	1.7	11.4	LC
Farina, quick-cooking	3½ oz.	43	1.3	8.9	LC
Farina, unenriched, regular	3½ oz.	42	1.3	8.7	LC
Fish flour from fillets	3½ oz.	398	93.0	0	HP
Fish flour from fillet waste	3½ oz.	305	71.0	0	HP
Fish flour from whole fish	3½ oz.	336	78.0	0	HP
Flako Muffin Mix	1 mfn.	133	2.7	20.7	HC

Food	Quantity	Calories	Protein Grams	Carbo-hydrates Grams	P/C/C Computer
Flako Popover Mix	1 popover	163	7	22.7	HC
Franco-American Macaroni and Cheese	7¼ oz.	200	8	26	HC
Franco-American Spaghetti in Tomato Sauce	7½ oz.	180	5	34	HC
Franco-American Spaghetti O's in Tomato Sauce	7½ oz.	170	5	32	HC
French or Vienna bread	3½ oz.	290	9.1	55.4	HCR
General Mills Baron Von Redberry	1 cup	108	1.5	24.7	HCR
General Mills Buck Wheats	1 cup	102	2.9	23.1	HCR
General Mills Cheerios	1¼ cup	112	3.8	20.2	HCR
General Mills Cocoa Puffs	1 cup	109	1.5	25.2	HCR
General Mills Count Chocula	1 cup	106	1.5	24.5	HCR
General Mills Country Corn Flakes	1¼ cup	111	2.4	24.3	HCR
General Mills Frankenberry	1 cup	109	1.3	24.9	HCR
General Mills Frosty O's	1 cup	112	1.8	24.2	HCR
General Mills Kaboom	1 cup	108	1.6	24.6	HCR
General Mills Kix	1½ cup	112	2.5	23.8	HCR
General Mills Lucky Charms	1 cup	110	2.1	23.5	HCR
General Mills Sir Grapefellow	1 cup	108	1.5	24.7	HCR
General Mills Sugar Jets	1 cup	111	2.1	23.7	HCR
General Mills Total	1¼ cup	100	2.5	23	HCR
General Mills Total Corn	1¼ cup	111	2.4	24.3	HCR
General Mills Trix	1 cup	110	1.5	25.2	HCR
General Mills Tutti-Frutti Twinkles	1 cup	112	1.9	24.2	HCR
General Mills Wheaties	1¼ cup	101	2.5	23	HCR
Green Giant Boil-in-Bag Frozen Brown Rice in Beef Stock	3½ oz.	120	2.5	20	LC
Green Giant Boil-in-Bag Frozen Rice Medley	3½ oz.	90	2	15	LC
Green Giant Boil-in-Bag Frozen Rice Pilaf	3½ oz.	90	1.8	18	LC
Green Giant Boil-in-Bag Frozen Rice Verdi	3½ oz.	100	1.8	20	LC
Green Giant Boil-in-Bag Frozen Spanish Rice	3½ oz.	80	1.7	16	LC
Green Giant Boil-in-Bag Frozen White and Wild Rice	3½ oz.	90	2.2	18	LC
H-O Enriched Farina	1 oz.	103	3.86	21.9	LP
H-O Instant Oatmeal	1 oz.	108	4.16	19.1	LP
H-O Old-Fashioned Oats	1 oz.	107	4.6	19.1	LP
H-O Quick Oats	1 oz.	105	4	18.6	LP
Hamburger Helper Chili Tomato	1/5 pkg.	160	4	34	HCR

Food	Quantity	Calories	Protein Grams	Carbo-hydrates Grams	P/C/C Computer
Heinz Macaroni in Cheese Sauce	3½ oz.	98	4.2	12.9	LP
Heinz Spaghetti and Tomato Sauce, Cheese	3½ oz.	72	2.2	13.1	LP
Heinz Spanish Rice	3½ oz.	74	1.3	14.1	LP
Italian bread	3½ oz.	276	9.1	56.4	HC
Johnnycake (northern style corn bread)	3½ oz.	267	8.7	45.5	HC
Kellogg's All-Bran	½ cup	100	3.4	20	HCR
Kellogg's Apple Jacks	1 cup	112	1.3	25.7	HCR
Kellogg's Bran Flakes, 40%	¾ cup	105	2.8	21.9	HCR
Kellogg's Cocoa Hoots	1 cup	110	1.3	25.6	HCR
Kellogg's Cocoa Krispies	1 cup	111	1.3	25.1	HCR
Kellogg's Concentrate	1/3 cup	108	11.3	15.2	HP
Kellogg's Corn Flakes	1-1/3 cup	108	2.1	24.2	HCR
Kellogg's Froot Loops	1 cup	116	1.7	24.7	HCR
Kellogg's Miniatures, Shredded Wheat	¾ cup	104	3.6	21.5	HCR
Kellogg's Pep	1 cup	108	2.7	23	HCR
Kellogg's Product 19	1 cup	104	2.4	23	HCR
Kellogg's Puffa Puffa Rice	1 cup	124	0.9	23.7	HCR
Kellogg's Puffed Rice	1 cup	55	1.0	12.2	LCR
Kellogg's Puffed Wheat	1 cup	51	2.1	10	LCR
Kellogg's Raisin Bran	½ cup	101	2.3	21.7	HCR
Kellogg's Rice Krispies	1 cup	109	1.8	24.5	HCR
Kellogg's Special K	1¼ cup	109	5.7	20.8	HCR
Kellogg's Sugar Frosted Flakes	¾ cup	109	1.4	25.1	HCR
Kellogg's Sugar Pops	1 cup	112	1.6	25.5	HCR
Kellogg's Sugar Smacks	1 cup	111	1.9	25	HCR
La Choy Fried Rice	2/3 cup	130.70	2.7	27.81	HCR
Lawry's Taco Shells	4-7/8 oz.	620.5	9.8	64	HC
Macaroni, enriched, cooked	3½ oz.	111	3.4	23.0	HCR
Macaroni, unenriched, cooked	3½ oz.	111	3.4	23.0	HCR
Macaroni and cheese, canned	3½ oz.	95	3.9	10.7	LC
Macaroni and cheese from home recipe	3½ oz.	215	8.4	20.1	HCR
Malt-o-Meal Chocolate Cereal	1 oz.	102	3	21.7	HCR
Malt-o-Meal Puffed Rice	½ oz.	50	1.0	10	LC
Malt-o-Meal Puffed Wheat	½ oz.	50	2	10	LC
Malt-o-Meal Quick Cereal	1 oz.	102	2.9	22.1	HCR
Millet, whole grain	3½ oz.	327	9.9	72.9	HC
Minute Rice Drumstick Packaged Rice Mix	½ cup	152	3.1	22.4	HCR
Minute Rice Rib Roast Packaged Rice Mix	½ cup	149	3.7	24.2	HCR
Minute Rice Spanish Packaged Rice Mix	½ cup	132	2.8	23	HCR
Minute Rice w/o Butter	½ cup	93	1.5	19.9	LC

Food	Quantity	Calories	Protein Grams	Carbo-hydrates Grams	P/C/C Com-puter
Monk's Bread Muffin	1 mfn.	129	5.42	23.8	HCR
Monk's White Bread	2 slices	156	5.56	28	HCR
Morton Blueberry Muffins	1 mfn.	115.6	2.4	20.8	HCR
Morton Brown 'n Serve Rolls	1 oz.	87.6	2.02	15.2	LC
Morton Corn Muffins	1 mfn.	132.4	2.81	20.4	HCR
Morton Parkerhouse Rolls	1 oz.	85.9	2.3	14.2	LC
Mrs. Wright's Break-Away Brown 'n Serve Rolls	1 oz.	80.6	2.17	14	LC
Mrs. Wright's Enriched White Bread	3½ oz.	256	7.7	49	HCR
Mrs. Wright's Hamburger Buns	3½ oz.	278	8.4	51.3	HCR
Mrs. Wright's Hot Dog Buns	3½ oz.	277	8.3	51	HCR
Mrs. Wright's Grecian Style Bread	3½ oz.	250	8.4	47	HCR
Mrs. Wright's Low Sodium Bread	3½ oz.	233	6.9	45.8	HCR
Mrs. Wright's Old-Fashioned, All Butter Bread	3½ oz.	248	7.3	48	HC
Mrs. Wright's Old-Fashioned Bread	3½ oz.	267	7.6	50	HC
Mrs. Wright's Old-Fashioned Sandwich Bread	3½ oz.	260.8	8.25	48	HC
Mrs. Wright's Super Soft Enriched White Bread	3½ oz.	259	7.8	49	HCR
Muffins, blueberry	3½ oz.	281	7.3	41.9	HCR
Muffins, bran	3½ oz.	261	7.7	43.1	HCR
Muffins, enriched degerminated cornmeal	3½ oz.	314	7.1	48.1	HC
Muffins, enriched flour	3½ oz.	294	7.8	42.3	HCR
Muffins, made from mixes	3½ oz.	324	6.9	50.0	HCR
Muffins, whole-ground cornmeal	3½ oz.	288	7.2	42.5	HC
Nabisco Bran Flakes, 100%	1 oz.	96.5	3.1	18.5	LC
Nabisco Cream of Wheat	1 oz.	101.6	3.03	21.6	LC
Nabisco Great Honey Crunchers Rice Cereal	1 oz.	113	0.96	24.7	LC
Nabisco Great Honey Crunchers Wheat	1 oz.	114.6	1.9	23.7	LC
Nabisco Instant Cream of Wheat	1 oz.	98.5	2.97	21	LC
Nabisco Miniatures, Shredded Wheat	1 oz.	107	2.97	22.5	LC
Nabisco Mix 'n Eat Cream of Wheat, Apple/Cinnamon	1 oz.	103	2.2	22	LC
Nabisco Mix 'n Eat Cream of Wheat, Maple/Brown Sugar	1 oz.	102	2.3	24	LC
Nabisco Quick Cream of Wheat	1 oz.	98.5	2.97	21	LC
Nabisco Shredded Wheat	1 oz.	99.3	2.8	21.6	LC
Nabisco Team Flakes	1 oz.	106.7	1.67	24.2	LC
Noodles, chow mein, canned	3½ oz.	489	13.2	58.0	HC

Food	Quantity	Calories	Protein Grams	Carbo-hydrates Grams	P/C/C Com-puter
Noodles, egg noodles	3½ oz.	125	4.1	23.3	HCR
Oat cereal with toasted wheat germ-soy grits	3½ oz.	62	3.3	9.5	LC
Oat flakes, maple-flavored, instant-cooking	3½ oz.	69	2.6	13.0	LC
Oat granules, maple-flavored, quick-cooking	3½ oz.	60	2.3	11.4	LC
Oatmeal (rolled oats)	3½ oz.	55	2.0	9.7	LC
Oats, puffed	3½ oz.	397	11.9	75.2	HC
Oats, puffed, sugar-coated	3½ oz.	396	6.7	85.6	HC
Oats, shredded	3½ oz.	379	18.8	72.0	HP
Oat-wheat cereal	3½ oz.	65	2.6	12.1	LC
Oregon Freeze Dry Noodles and Chicken	3½ oz.	450	18.5	55.5	HCR
Oregon Freeze Dry Noodles Stroganoff	3½ oz.	490	20.5	47.5	HC
Ovenjoy Enriched White Bread	3½ oz.	258	8	49	HC
Pancakes, home-made with enriched flour	3½ oz.	231	7.1	34.1	HCR
Pancakes, home-made with unenriched flour	3½ oz.	231	7.1	34.1	HCR
Pancakes and waffles from mixes	3½ oz.	202	6.1	31.9	HCR
Pantry Pride Corn Flakes	1 oz.	110	2	24	HCR
Pantry Pride Cracked Wheat Bread	2 slices	150	5	33	HCR
Pantry Pride Enriched Flour	¼ cup	95	3	20	LC
Pantry Pride Frozen Waffles	1 section	55	1	10.5	LC
Pantry Pride Hamburger Buns	1 bun	106	3	19	LC
Pantry Pride Hot Dog Buns	1 bun	106	3	19	LC
Pantry Pride Macaroni	2 oz.	210	8	42	HC
Pantry Pride Sugar Frosted Flakes	1 oz.	110	1	25	HC
Pantry Pride White Bread	2 slices	150	5	27	HC
Pantry Pride White Rice	½ cup	118	2	26	HC
Pastinas:					
Egg	3½ oz.	383	12.9	71.8	HC
Carrot	3½ oz.	371	11.9	75.7	HC
Spinach	3½ oz.	368	12.4	74.8	HC
Peanut flour, defatted	3½ oz.	371	47.9	31.5	HP
Pepperidge Farm Butter Crescent Rolls	1 roll	135	2.5	13.2	HC
Pepperidge Farm Butterfly Rolls	1 roll	57	1.4	7.9	LC
Pepperidge Farm Cinnamon Raisin Bread	2 slices	149	3.2	27.1	HC
Pepperidge Farm Club Rolls	1 roll	120	4	23.3	HC
Pepperidge Farm Corn Bread Stuffing	3½ oz.	366	10.7	73.1	HCR

Food	Quantity	Calories	Protein Grams	Carbo-hydrates Grams	P/C/C Com-puter
Pepperidge Farm Corn and Molasses Bread	2 slices	143	4	28.7	HCR
Pepperidge Farm Cracked Wheat Bread	2 slices	139	4.7	25.6	HCR
Pepperidge Farm Cube Stuffing	3½ oz.	378	11.7	73.5	HCR
Pepperidge Farm Dinner Rolls	1 roll	80	1.6	9.9	LCR
Pepperidge Farm English Tea Bread	2 slices	144	4.3	23.9	HCR
Pepperidge Farm Finger Party Pan Rolls	1 roll	75	1.5	9.2	LCR
Pepperidge Farm French Bread	1 slice	87	2.5	15.8	HC
Pepperidge Farm Golden Twist Rolls	1 roll	131	2.4	14	HC
Pepperidge Farm Hamburger Buns	1 bun	112	3.3	18.9	HC
Pepperidge Farm Hearth Rolls	1 roll	64	1.9	11.6	LC
Pepperidge Farm Herb Stuffing	3½ oz.	366	11.2	78.2	HCR
Pepperidge Farm Honey Wheatberry Bread	2 slices	156	5.9	30.7	HCR
Pepperidge Farm Hot Dog Buns	1 bun	114	3.4	19.3	HC
Pepperidge Farm Italian Bread	1 slice	90	2.6	16.3	LC
Pepperidge Farm Large White Bread	2 slices	154	4.2	26.2	HC
Pepperidge Farm Oatmeal Bread	2 slices	133	4.6	24.2	HCR
Pepperidge Farm Old-Fashioned Rolls	1 roll	57	1.4	7.6	LC
Pepperidge Farm Party Rye Bread	4 slices	62	2.1	11.9	LC
Pepperidge Farm Pumpernickel Bread	2 slices	158	5.6	31.5	HCR
Pepperidge Farm Pumpernickel Party Bread	4 slices	79	2.7	15.8	LC
Pepperidge Farm Round Party Pan Rolls	1 roll	45	0.9	5.5	LC
Pepperidge Farm Rye Bread	2 slices	164	5.4	31.4	HC
Pepperidge Farm Sandwich White Bread	2 slices	141	3.7	22.7	HC
Pepperidge Farm Seedless Rye Bread	2 slices	164	5.4	30.9	HC
Pepperidge Farm Sprouted Rye Bread	2 slices	130	5.9	20.4	HC
Pepperidge Farm Sprouted Wheat Bread	2 slices	129	5.8	22.6	HC
Pepperidge Farm Toasting White Bread	2 slices	169	5.7	31.1	HCR
Pepperidge Farm Triple French Rolls	1 roll	264	8.7	51.4	HCR
Pepperidge Farm Twin French Rolls	1 roll	389	12.8	75.7	HC

Breakfast Foods, Breads, Cereals, Flours, Grains

Food	Quantity	Calories	Protein Grams	Carbo-hydrates Grams	P/C/C Com-puter
Pepperidge Farm Whole Wheat Bread	2 slices	131	5.3	22.7	HC
Pillsbury Apricot Nut Sweet Bread	1/12 loaf	180	3	30	HC
Pillsbury Ballard Buttermilk Biscuit Mix	2 bis.	120	3	22	HC
Pillsbury Ballard Corn Bread Mix	1 piece	80	2	13	LC
Pillsbury Ballard Crescent Rolls	1 roll	80	2	13	LC
Pillsbury Banana Sweet Bread	1/12 loaf	160	3	29	HCR
Pillsbury Blueberry Nut Sweet Bread	1/12 loaf	150	2	26	HCR
Pillsbury Butterflake Rolls	1 roll	55	1.5	8.5	LC
Pillsbury Buttermelt Biscuit Mix	2 bis.	120	2	18	HC
Pillsbury Buttermilk Biscuit Mix	2 bis.	120	3	22	HC
Pillsbury Cherry Nut Sweet Bread	1/12 loaf	180	3	31	HCR
Pillsbury Country Style Biscuit Mix	2 bis.	120	3	22	HC
Pillsbury Cranberry Sweet Bread	1/12 loaf	170	3	31	HCR
Pillsbury Crescent Rolls	1 roll	95	1.5	13.5	LC
Pillsbury Date Sweet Bread	1/12 loaf	170	3	33	HC
Pillsbury 1869 Biscuit Mix	2 bis.	160	3	18	HC
Pillsbury 1869 Butter Tastin' Biscuit Mix	2 bis.	230	3	27	HC
Pillsbury 1869 Buttermilk Biscuit Mix	2 bis.	160	3	18	HC
Pillsbury 1869 Heat 'n Serve Biscuit Mix	2 bis.	200	3	24	HC
Pillsbury Extra Light Buttermilk Biscuit Mix	2 bis.	110	3	21	HCR
Pillsbury Extra Light Pancake Mix	3-4 pan.	270	7	35	HC
Pillsbury Farina w/Milk and Salt	¾ cup	230	9	32	HC
Pillsbury Farina w/Water and Salt	¾ cup	100	3	23	HCR
Pillsbury Granola	¼ cup	180	6	23	HC
Pillsbury Granola w/Coconut and Cashews	¼ cup	190	6	21	HC
Pillsbury Granola w/Raisins and Almonds	¼ cup	180	6	23	HC
Pillsbury Heat 'n Serve Buttermilk Biscuit Mix	2 bis.	190	3	24	HC

Food	Quantity	Calories	Protein Grams	Carbo-hydrates Grams	P/C/C Com-puter
Pillsbury Hungry Jack Blueberry Pancake Mix	3-4 pan.	360	8	48	HC
Pillsbury Hungry Jack Butter Tastin' Biscuit Mix	2 bis.	180	3	22	HC
Pillsbury Hungry Jack Butter-milk Biscuit Mix	2 bis.	120	3	21	HC
Pillsbury Hungry Jack Butter-milk Crescent Rolls	1 roll	115	2	15	HC
Pillsbury Hungry Jack Butter-milk Pancake and Waffle Mix	3-4 pan.	240	7	29	HC
Pillsbury Hungry Jack Com-plete Buttermilk Pancake Mix	3-4 pan.	220	5	42	HCR
Pillsbury Hungry Jack Com-plete Pancake Mix	3-4 pan.	220	5	42	HCR
Pillsbury Hungry Jack Corn Bread Mix	1 piece	80	1.5	12.5	LC
Pillsbury Hungry Jack Extra Light Pancake Mix	3-4 pan.	220	7	32	HCR
Pillsbury Hungry Jack Flaky Biscuit Mix	2 bis.	160	4	23	HC
Pillsbury Hungry Jack Flaky Buttermilk Biscuit Mix	2 bis.	170	4	24	HC
Pillsbury Hungry Jack Fluffy Biscuit Mix	2 bis.	180	3	22	HC
Pillsbury Hungry Jack Oven Ready Biscuit Mix	2 bis.	120	3	22	HC
Pillsbury Hungry Jack Sweet Corn Bread Mix	1 piece	100	1.5	13	LC
Pillsbury Italian Crescent Rolls	1 roll	90	1.5	11.5	LC
Pillsbury Nut Sweet Bread	1/12 loaf	190	3	29	HC
Pillsbury Pan Rolls	1 roll	75	1.5	11.5	LC
Pillsbury Parkerhouse Rolls	1 roll	60	1.5	11	LC
Pillsbury Snoflake Dinner Rolls	1 roll	70	2	11.5	LC
Pillsbury Tenderflake Biscuit Mix	2 bis.	120	2	16	HC
Pillsbury Tenderflake Butter-milk Biscuit Mix	2 bis.	120	2	16	HC
Pillsbury Toasted Wheat Germ	¼ cup	120	9	13	HC
Pizza, with cheese, home recipe, baked:					
Chilled-baked	3½ oz.	245	9.2	36.3	HC
Frozen-baked	3½ oz.	245	9.5	35.4	HC
With cheese topping	3½ oz.	236	12.0	28.3	HC
With sausage topping	3½ oz.	234	7.8	29.6	HC

Breakfast Foods, Breads, Cereals, Flours, Grains

Food	Quantity	Calories	Protein Grams	Carbo-hydrates Grams	P/C/C Com-puter
Popovers, home-baked with enriched flour	3½ oz.	224	8.8	25.8	HC
Post Alpha-Bits	1 cup	113	2.2	23	HCR
Post Bran Flakes, 40%	2/3 cup	97	2.5	21	LC
Post Cinnamon Raisin Bran	½ cup	92	2	21	LC
Post Cocoa Pebbles	7/8 cup	111	0.9	25	HCR
Post Crispy Critters	1 cup	113	2.2	23	HCR
Post Fortified Oat Flakes	2/3 cup	107	5.1	19	LC
Post Frosted Rice Krinkies	7/8 cup	111	0.9	25	HCR
Post Fruity Pebbles	7/8 cup	111	0.9	25	HCR
Post Grape Nuts	¼ cup	104	2.5	23	HCR
Post Grape Nuts Flakes	2/3 cup	101	2.5	22	HCR
Post Honeycomb Sweet Crisp Corn	1-1/3 cup	108	1.6	25	HCR
Post Pink Panther Flakes	2/3 cup	112	1.0	25	HCR
Post Raisin Bran	½ cup	92	2	21	LC
Post Super Orange Crisp Wheat Puffs	1 cup	109	1.6	25	HCR
Post Super Sugar Crisp Wheat Puffs	7/8 cup	107	1.7	25	HCR
Post Toasties	1 cup	108	2	24	HCR
Potato flour	3½ oz.	351	8.0	79.9	HCR
Presto Self-Rising Flour	¼ cup	100	2	21	LC
Prince Egg Noodles	2 oz.	220	8	40	HC
Prince Macaroni	2 oz.	210	8	42	HC
Prince Macaroni in Cheese Sauce	1 cup	253.33	10.67	45.33	HC
Prince Spaghetti	2 oz.	210	8	42	HC
Pumpernickel bread	3½ oz.	246	9.1	53.1	HC
Quaker Cap'n Crunch	¾ cup	123	1.3	22.8	HC
Quaker Cap'n Crunch Crunchberries	¾ cup	119	1.3	23.8	HC
Quaker Cap'n Crunch Peanut Butter	¾ cup	128	2.1	21.3	HC
Quaker Cap'n Crunch Vanilly Crunch	¾ cup	116	1.3	23.8	HC
Quaker Enriched Egg Noodles	2 oz.	207	7.62	40.4	HCR
Quaker Enriched Elbow Macaroni	2 oz.	205	7.09	42.6	HCR
Quaker Enriched Farina	1/6 cup	100	2.5	22.1	HCR
Quaker Enriched Spaghetti	2 oz.	209	7.09	43.2	HCR
Quaker Instant Oatmeal	1 oz.	107	3.8	19	LCR
Quaker Instant Oatmeal w/Apples	1 oz.	105.9	2.6	21.4	HC
Quaker King Vitamin	¾ cup	118	1.1	24	HC
Quaker Life	2/3 cup	107	5.1	20.4	LC
Quaker Old-Fashioned Oats	1/3 cup	107	4.1	18.8	LC
Quaker Oatmeal w/Maple and Brown Sugar	1 pkg.	177	3.9	36.2	HCR
Quaker Oatmeal w/Raisins and Spice	1 pkg.	154	3.6	32.1	HCR

Food	Quantity	Calories	Protein Grams	Carbo-hydrates Grams	P/C/C Com-puter
Quaker 100% Natural Cereal	¼ cup	140	3.4	17	HC
Quaker 100% Natural Cereal w/Fruit	¼ cup	136	2.8	17.8	HC
Quaker Puffed Rice	1 cup	56	0.9	12.5	LC
Quaker Puffed Wheat	1 cup	51	1.9	11.2	LC
Quaker Quisp	1-1/6 cup	122	1.3	23.1	HCR
Quaker Shredded Wheat	1 oz.	101.3	2.7	21.5	HC
Raisin bread	3½ oz.	262	6.6	53.6	HCR
Ralston Purina Corn Chex	1 oz.	108	2	24	HCR
Ralston Purina Instant Ralston	1 oz.	106	3.4	20.7	LC
Ralston Purina Ralston	1 oz.	106	3.4	20.7	LC
Ralston Purina Rice Chex	1 oz.	112	1.3	24.5	HCR
Ralston Purina Super Sugar Chex	1 oz.	125	1.0	23	HCR
Ralston Purina Wheat Chex	1 oz.	112	3	22.5	HCR
Red Cross Enriched Egg Noodles	2 oz.	220	8	40	HCR
Red Cross Enriched Macaroni	2 oz.	210	7	41	HCR
Red Skillet Egg Noodles and Burgundy Sauce Dinner	8 oz.	190	12	30	HCR
Red Skillet Egg Noodles and Mushroom Sauce Dinner	8 oz.	180	12	28	HCR
Red Skillet Egg Noodles Stroganoff	8 oz.	220	13	29	HCR
Red Skillet Macaroni and Cheese Sauce	8 oz.	200	14	32	HCR
Rice:					
Brown	3½ oz.	119	2.5	25.5	HCR
White	3½ oz.	109	2.0	24.2	HCR
Rice bran	3½ oz.	276	13.3	50.8	HCR
Rice cereal product	3½ oz.	50	.8	11.2	LC
Rice cereal with protein concentrate	3½ oz.	382	40.0	54.8	HC
Rice flakes cereal	3½ oz.	390	6.0	87.7	HC
Rice polish	3½ oz.	265	12.1	57.7	HCR
Rice, puffed, cereal	3½ oz.	399	6.0	89.5	HC
Rice, shredded, cereal	3½ oz.	392	5.2	88.8	HC
Riceland Milled Rice	½ cup	118	2	26	HCR
Roll dough and rolls baked from dough	3½ oz.	311	8.5	56.0	HCR
Roll mix and rolls baked from mix	3½ oz.	299	9.0	54.5	HCR
Rolls and buns:					
Danish pastry	3½ oz.	422	7.4	45.6	HC
Hard rolls	3½ oz.	312	9.8	59.5	HCR
Plain pan rolls	3½ oz.	298	8.2	53.0	HCR
Raisin rolls or buns	3½ oz.	275	6.9	56.4	HCR
Sweet rolls	3½ oz.	316	8.5	49.3	HCR

Breakfast Foods, Breads, Cereals, Flours, Grains

Food	Quantity	Calories	Protein Grams	Carbo-hydrates Grams	P/C/C Com-puter
Roman Meal	2/3 cup	130	5	23	HC
Roman Meal, Instant	2/3 cup	130	5	23	HC
Roman Meal Brown 'n Serve Rolls	1 roll	65	3	13.5	LC
Roman Meal Hamburger Buns	1 bun	110	4	20	HC
Roman Meal Light Bread	2 slices	150	7	28	HCR
Roman Meal Refrigerated Biscuit Mix	2 bis.	200	5	31	HCR
Roman Meal Wafers	4 wafers	100	3	17	LC
Roman Meal White Bread	2 slices	140	6	27	HC
Ronzoni Enriched Egg Noddles	2 oz.	220	8	40	HCR
Ronzoni Enriched Macaroni	2 oz.	210	7	41	HC
Rusk	3½ oz.	419	13.8	71.0	HCR
Rye:					
Dark flour	3½ oz.	327	16.3	68.1	HCR
Light flour	3½ oz.	357	9.4	77.9	HCR
Medium flour	3½ oz.	350	11.4	74.8	HCR
Whole grain	3½ oz.	334	12.1	73.4	HCR
Rye bread	3½ oz.	243	9.1	52.1	HCR
Rye wafers, whole-grain	3½ oz.	344	13.0	76.3	HCR
Safeway Corn Flakes	1 oz.	110	2	24	HCR
Safeway Enriched Thin Sliced White Bread	3½ oz.	256	8.17	48	HCR
Safeway Enriched White Bread	3½ oz.	256	8.1	48	HCR
Safflower seed meal, partially defatted	3½ oz.	355	39.6	36.5	HP
Salt-rising bread	3½ oz.	267	7.9	52.2	HCR
Salt sticks:					
Regular type	3½ oz.	384	12.0	75.3	HCR
Vienna bread type	3½ oz.	304	9.5	58.0	HCR
Shake 'n Bake Coating Mix for Chicken	1 env.	276	7.2	40	HCR
Shake 'n Bake Coating Mix for Fish	1 env.	224	6.5	33	HC
Shake 'n Bake Coating Mix for Hamburger	1 env.	158	7.2	31.6	HC
Shake 'n Bake Coating Mix for Pork	1 env.	260	7.2	46.3	HCR
Skylark Brown 'n Serve Rolls	1 oz.	68.2	1.96	14	LC
Skylark Buttermilk Bread	3½ oz.	258	8.5	48	HCR
Skylark Buttermilk Brown 'n Serve Rolls	1 oz.	79.4	2.46	14.3	LC
Skylark Cloverleaf Brown 'n Serve Rolls	1 oz.	81.7	2.2	14.6	LC
Skylark Egg Hamburger Buns	3½ oz.	274	8	51	HCR
Skylark Egg Rolls	1 oz.	78.3	2.28	14.6	LC
Skylark Enriched White Bread	3½ oz.	255	8	49	HC

Food	Quantity	Calories	Protein Grams	Carbo- hydrates Grams	P/C/C Com- puter
Skylark Flaky Gem Brown 'n Serve Rolls	1 oz.	81.7	2.2	14.6	LC
Skylark French Bread	3½ oz.	243	7.5	47	HC
Skylark Hamburger Buns	3½ oz.	280	8.5	51	HCR
Skylark Hot Dog Buns	3½ oz.	280	8.5	51	HCR
Skylark Italian Bread	3½ oz.	244	7.67	50	HCR
Skylark Italian Formula Bread	3½ oz.	260	7.6	47	HCR
Skylark 100% Butter Bread	3½ oz.	257	8.1	49	HCR
Skylark Parkerhouse Rolls	1 oz.	80.6	2.4	14.3	LC
Skylark Sesame Brown 'n Serve Rolls	1 oz.	82.6	2.16	14.4	LC
Skylark Sesame Hamburger Buns	3½ oz.	282	9.2	51.1	HCR
Skylark Sesame Rolls	1 oz.	80.6	2.5	13.7	LC
Skylark Sesame Seed Bread	3½ oz.	251	8	48	HC
Skylark Sesame Twist Bread	3½ oz.	261	8.75	48	HC
Skylark Sour French Bread	3½ oz.	256	9.23	49	HC
Skylark Split Top Bread	3½ oz.	256	8	48	HCR
Skylark Stonehedge Farms Bread	3½ oz.	254	9	47	HCR
Skylark Sweet Farmstyle Rolls	1 oz.	92.6	2.64	16.8	LC
Skylark Tea Rolls	1 oz.	79.4	2.4	14.5	LC
Skylark Twin Brown 'n Serve Rolls	1 oz.	81.7	2.2	14.6	LC
Skylark Vienna Style Bread	3½ oz.	251	7.75	49	HCR
Skylark Western Farms Bread	3½ oz.	255	8.33	48	HCR
Sorghum grain, all types	3½ oz.	332	11.0	73.0	HCR
Soybean flours:					
Defatted	3½ oz.	326	47.0	38.1	HP
Full fat	3½ oz.	421	36.7	30.4	HP
High fat	3½ oz.	380	41.2	33.3	HP
Low fat	3½ oz.	356	43.4	36.6	HP
Spaghetti	3½ oz.	148	5.0	30.1	HC
Spaghetti in tomato sauce with cheese:					
Canned	3½ oz.	76	2.2	15.4	LC
Home recipe	3½ oz.	104	3.5	14.8	HC
Spanish rice, home recipe	3½ oz.	87	1.8	16.6	LC
Spark-o-Life Enriched White Bread	3½ oz.	257	9.49	48	HC
Spoonbread	3½ oz.	195	6.7	16.9	HC
Sun Country Granola	2 oz.	250	9	34	HCR
Sun Country Honey Almond Granola	2 oz.	250	9	34	HCR
Sunflower seed flour, partially defatted	3½ oz.	339	45.2	37.7	HC
Swanson French Toast w/Sausage Frozen Breakfast	4½ oz.	300	16	22	HC

Food	Quantity	Calories	Protein Grams	Carbo-hydrates Grams	P/C/C Computer
Swanson Pancakes w/Sausage Frozen Breakfast	6 oz.	480	18	44	HCR
Swanson Scrambled Eggs/Sausage/ Cake Frozen Breakfast	6¼ oz.	420	12	22	HC
Tillie Lewis Tasti-Diet Butter-milk Pancake Mix	4 cakes	42	2	8	LC
Tillie Lewis Tasti-Diet Regular Pancake Mix	4 cakes	42	2	8	LC
Torula Yeast	3½ oz.	277	38.6	37.0	HP
Town House Enriched Egg Noodles	2 oz.	220	8	40	HCR
Town House Instant Rice	½ cup	82.5	1.5	18.8	LP
Town House Quick Oats	1 oz.	110	4	20	HC
Town House Regular Oats	1 oz.	110	4	20	HC
Town House Spaghetti	2 oz.	210	7	41	HC
Town House White Enriched Rice	½ cup	115	2	26	HC
Uncle Ben's Beef-Flavored Packaged Rice Mix w/Butter	½ cup	120	2.8	21.7	HC
Uncle Ben's Beef-Flavored Packaged Rice Mix w/o Butter	½ cup	103	2.8	21.6	HC
Uncle Ben's Brown Rice w/Butter	½ cup	114.8	2.2	21.5	HC
Uncle Ben's Brown Rice w/o Butter	½ cup	100	2.2	21.4	HC
Uncle Ben's Brown and Wild Rice w/Butter	½ cup	117	2.7	20.3	HC
Uncle Ben's Brown and Wild Rice w/o Butter	½ cup	99	2.7	20.3	HC
Uncle Ben's Chicken-Flavored Packaged Rice Mix w/Butter	½ cup	133	2.7	20.6	HC
Uncle Ben's Chicken-Flavored Packaged Rice Mix w/o Butter	½ cup	100	2.6	20.5	HC
Uncle Ben's Curried Packaged Rice Mix w/Butter	½ cup	117	2.8	22	HC
Uncle Ben's Curried Packaged Rice Mix w/o Butter	½ cup	100	2.8	21.9	HC
Uncle Ben's Herb Stuffing	3½ oz.	263	12.9	74.3	HCR
Uncle Ben's Instant Rice w/Butter	½ cup	94.5	1.4	18.2	LC
Uncle Ben's Instant Rice w/o Butter	½ cup	78.8	1.4	18.1	LC
Uncle Ben's Long Grain Rice w/o Butter	½ cup	90.7	2.3	27.6	HCR
Uncle Ben's Long Grain and Wild Rice w/Butter	½ cup	114	3	21	HC
Uncle Ben's Long Grain and Wild Rice w/o Butter	½ cup	97	3	21	LC

Food	Quantity	Calories	Protein Grams	Carbo-hydrates Grams	P/C/C Com-puter
Uncle Ben's Pilaf w/Butter Packaged Rice Mix	½ cup	129	2.4	21.1	HC
Uncle Ben's Pilaf w/o Butter Packaged Rice Mix	½ cup	97	2.3	21.1	LC
Uncle Ben's Spanish Packaged Rice Mix w/Butter	½ cup	129	3.8	23.3	HC
Uncle Ben's Spanish Packaged Rice Mix w/o Butter	½ cup	109	3.8	23.3	HC
Van Camp Spaghetti in Tomato Sauce	1 cup	168	4.8	33.8	HCR
Waffles, Home-baked	3½ oz.	279	9.3	37.5	HCR
Waffle mixes and waffles baked from mixes	3½ oz.	305	4.8	40.2	HCR
Wheat bran	3½ oz.	213	16.0	61.9	HCR
Wheat flakes	3½ oz.	354	10.2	80.5	HCR
Wheat flours	all 3½ oz.				
80% extraction from hard wheats		365	12	74.1	HCR
Patent:					
All-purpose or family flour		364	10.5	76.1	HCR
Bread flour		365	11.8	74.7	HCR
Cake or pastry flour		364	7.5	79.4	HCR
Gluten flour		378	41.4	47.2	HCR
Self-rising flour		352	9.3	74.2	HCR
Straight, hard wheat		365	11.8	74.5	HCR
Straight, soft wheat		364	9.7	76.9	HCR
Whole, from hard wheats		333	13.3	71.0	HCR
Wheat germ	3½ oz.	363	26.6	46.7	HCR
Wheat germ, toasted	3½ oz.	391	30.0	49.5	HCR
Wheat and malted barley cereal, toasted	3½ oz.	65	2.0	13.2	LC
Wheat and malted barley flakes	3½ oz.	392	8.8	84.3	HCR
Wheat and malted barley granules	3½ oz.	391	10.0	84.8	HCR
Wheat, puffed	3½ oz.	363	15.0	78.5	HCR
Wheat, puffed, with sugar-honey	3½ oz.	376	6.0	88.3	HCR
Wheat, rolled, hot breakfast cereal	3½ oz.	75	2.2	16.9	LC
Wheat, shredded	3½ oz.	354	9.9	79.9	HCR
Wheat, shredded, with malt-salt-sugar	3½ oz.	366	9.1	81.7	HCR
Wheat, whole grain	all 3½ oz.				
Durum		332	12.7	70.1	HCR
Hard red spring		330	14.0	69.1	HCR

Breakfast Foods, Breads, Cereals, Flours, Grains

Food	Quantity	Calories	Protein Grams	Carbo- hydrates Grams	P/C/C Com- puter
Wheat, whole grain *(cont.)*:					
Hard red winter		330	12.3	71.7	HCR
Soft red winter		326	10.2	72.1	HCR
White		335	9.4	75.4	HCR
Wheat, whole-meal, hot break- fast cereal	3½ oz.	45	1.8	9.4	LC
White bread	3½ oz.	269	8.7	50.4	HC
Whole-wheat bread	3½ oz.	243	10.5	47.7	HCR
Whole-wheat rolls	3½ oz.	257	10.0	52.3	HCR
Wild rice	3½ oz.	353	14.1	75.3	HCR
Wonder Beefsteak Rye Bread	2 slices	157.9	6.3	31.1	HC
Wonder Brown 'n Serve Rolls	1 roll	78.8	2.12	15.9	HC
Wonder Buttertop Wheat Bread	2 slices	121.8	4.84	23	HC
Wonder Buttertop White Bread	2 slices	139	4.2	26.3	HC
Wonder Cinnamon Raisin Bread	2 slices	120.8	3.4	26.1	HC
Wonder Cracked Honey Wheat Bread	2 slices	125.1	3.98	22.5	HC
Wonder Daffodil White Bread	2 slices	116.7	4.3	22.6	HC
Wonder English Muffins	1 mfn.	136.9	4.3	26.7	HC
Wonder Golden Wheat Bread	2 slices	126.8	4.7	24.2	HC
Wonder Hamburger Buns	1 bun	122	3.5	21.3	HC
Wonder Hot Dog Buns	1 bun	122	3.5	21.3	HC
Wonder Profile Dark Rye Bread	2 slices	110	5.3	22	HC
Wonder Profile Light Rye Bread	2 slices	113.7	4.82	22.5	HC
Wonder Raisin Round Muffins	1 round	148.5	4.36	30.3	HCR
Wonder Rye Bread	2 slices	109.1	4.6	21	HC
Wonder Scone Muffin	1 scone	136.9	4.02	27.9	HC
Wonder White Bread	2 slices	131.3	4.1	25.3	HC
Wonder Whole Wheat Bread	2 slices	118.2	4.46	22.7	HC
Yeast, Baker's:					
Compressed	3½ oz.	86	12.1	11.0	HP
Dry (active)	3½ oz.	282	36.9	38.9	HP

Fruits
and Fruit Products

Food	Quantity	Calories	Protein Grams	Carbo-hydrates Grams	P/C/C Com-puter
Acerola berry	3½ oz.	28	.4	6.8	LC
Apple, raw	3½ oz.	58	.2	14.5	LC
Apple butter	3½ oz.	186	.4	46.8	HCR
Applesauce, sweetened	3½ oz.	91	.2	23.8	LC
Apricot nectar	3½ oz.	57	.3	14.6	LC
Apricots, candied	3½ oz.	338	.6	86.5	HC
Apricots, canned, heavy syrup	3½ oz.	86	.6	22.0	HCR
Apricots, canned, water pack	3½ oz.	38	.7	9.6	LC
Apricots, dried	3½ oz.	260	5.0	66.5	HCR
Apricots, raw	3½ oz.	51	1.0	12.8	LC
Avocado	3½ oz.	167	2.1	6.3	LC
Banana	3½ oz.	85	1.1	22.2	LC
Banana, dehydrated or banana powder	3½ oz.	340	4.4	88.6	HCR
Banana, red	3½ oz.	90	1.2	23.4	LC
Birds Eye Frozen Strawberry Halves	½ cup	162	0.9	39.9	HCR
Birds Eye Frozen Whole Strawberries	¼ pkg.	101	0.6	26.6	LC
Birds Eye Quick Thaw Blueberries	½ cup	114	0.7	28.7	HCR
Birds Eye Quick Thaw Mixed Fruit	½ cup	111	0.7	28.4	HCR
Birds Eye Quick Thaw Raspberries	½ cup	148	1.1	37.2	HCR
Birds Eye Quick Thaw Rhubarb	½ cup	84	0.6	21.4	LC
Birds Eye Quick Thaw Sliced Peaches	½ cup	87	0.7	22.3	LC
Birds Eye Quick Thaw Strawberries	½ cup	122	0.7	30.7	HCR
Birds Eye Quick Thaw Sweet Cherries	½ cup	122	1.4	30.8	HCR
Blackberries, canned	3½ oz.	54	.8	12.1	LC
Blackberries, raw	3½ oz.	58	1.2	12.9	LC
Blueberries:					
Canned	3½ oz.	39	.5	9.8	LC
Frozen	3½ oz.	55	.7	13.6	LC
Raw	3½ oz	62	.7	15.3	LC
Breadfruit, raw	3½ oz.	103	1.7	26.2	HC
Cantaloupe	3½ oz.	30	.7	7.5	LC
Carissa (natal plum)	3½ oz.	70	.5	16.0	LC
Casaba melon (Golden Beauty)	3½ oz.	27	1.2	6.5	LC
Cherries:					
Candied	3½ oz.	339	.5	86.7	HC
Maraschino, bottled	3½ oz.	116	.2	29.4	HC
Raw, sour, red	3½ oz.	58	1.2	14.3	LC
Raw, sweet	3½ oz.	70	1.3	17.4	LC

Food	Quantity	Calories	Protein Grams	Carbo-hydrates Grams	P/C/C Computer
Cherries *(cont.):*					
Sour, red, canned	3½ oz.	74	.8	18.7	LC
Sour, red, frozen	3½ oz.	112	1.0	27.8	HC
Crab apples	3½ oz.	68	.4	17.8	LC
Cranberries	3½ oz.	46	.4	10.8	LC
Cranberry sauce, canned, strained	3½ oz.	146	.1	37.5	HC
Cranberry sauce, home-prepared, unstrained	3½ oz.	178	.2	45.5	HC
Cranberry-orange relish	3½ oz.	178	.4	45.4	HC
Currants, red, black, and white	3½ oz.	54	1.7	13.1	LC
Dates, domestic, natural, and dry	3½ oz.	274	2.2	72.9	HCR
Del Monte Apricots	½ cup	104.5	0.9	27.5	HC
Del Monte Cling Peaches	½ cup	95	0.65	25.1	HC
Del Monte Dark Cherries	½ cup	91.5	0.85	23	HC
Del Monte Dried Apples	3½ oz.	93	0.26	23.9	HC
Del Monte Dried Apricots	3½ oz.	135	2.0	34.3	HCR
Del Monte Dried Peaches	3½ oz.	174	2.4	44.6	HCR
Del Monte Dried Pears	3½ oz.	156	1.5	40.6	HCR
Del Monte Dried Prunes	3½ oz.	207	2.1	51.4	HCR
Del Monte Figs	½ cup	104	0.6	27.4	HC
Del Monte Freestone Peaches	½ cup	108	0.5	29.1	HC
Del Monte Fruit Cocktail	½ cup	87.5	0.45	23.5	LC
Del Monte Fruit Cup, Mixed Fruit	5 oz.	100	0.4	27.4	LC
Del Monte Fruit Cup, Peaches	5¼ oz.	107	0.4	29.5	LC
Del Monte Fruit Salad	½ cup	110	0.65	29.3	HC
Del Monte Fruits for Salad	½ cup	78.5	0.35	21.4	LC
Del Monte Golden Seedless Raisins	3½ oz.	183	2.1	48.4	HCR
Del Monte Grapefruit Sections in Juice	½ cup	45	0.75	10.4	LC
Del Monte Grapefruit Sections in Syrup	½ cup	42.3	0.25	18.6	LC
Del Monte Mandarin Oranges	½ cup	77	0.65	20.6	LC
Del Monte Mixed Dried Fruit	3½ oz.	193	1.7	51.7	HC
Del Monte Pears	½ cup	84	0.35	21.7	LC
Del Monte Pineapple in Juice	½ cup	79	0.7	20.7	LC
Del Monte Pitted Prunes	½ cup	144	1.15	37.5	HCR
Del Monte Prunes in Moist-Pak	½ cup	269	3.5	70.3	HCR
Del Monte Pumpkin	½ cup	39	0.85	9.6	LC
Del Monte Purple Plums	½ cup	112.5	0.6	30.4	HC
Del Monte Royal Anne Cherries	½ cup	109	1.05	28.1	HC
Del Monte Spiced Peaches	½ cup	94.5	0.5	29.5	HCR
Del Monte Stewed Prunes	½ cup	158	1.05	41.8	HC
Del Monte Sweetened Applesauce	½ cup	119	0.25	32.5	HC

Food	Quantity	Calories	Protein Grams	Carbo-hydrates Grams	P/C/C Com-puter
Del Monte Thompson Seedless Raisins	3½ oz.	186	2.0	48.5	HC
Del Monte Zante Currants-Raisins	3½ oz.	179	2.1	47.4	HC
Dole Pineapple Chunks in Syrup	½ cup	84	0.34	22	LC
Dole Pineapple Chunks in Water	½ cup	64	0.45	16.5	LC
Dromedary Chopped Dates	3½ oz.	350	1.9	81.1	HC
Dromedary Pitted Dates	3½ oz.	332	1.7	79.3	HC
Elderberries	3½ oz.	72	2.6	16.4	LC
Figs:					
Candied	3½ oz.	299	3.5	73.7	HCR
Canned	3½ oz.	84	.5	21.8	LC
Dried	3½ oz.	274	4.3	69.1	HC
Raw	3½ oz.	80	1.2	20.3	LC
Fruit cocktail, canned, heavy syrup	3½ oz.	76	.4	19.7	LC
Fruit cocktail, canned, water-packed	3½ oz.	37	.4	9.7	LC
Fruit salad, canned, heavy syrup	3½ oz.	75	.3	19.4	LC
Fruit salad, canned, water-packed	3½ oz.	35	.4	9.1	LC
Gelatin, dry	3½ oz.	335	85.6	0	HP
Gelatin dessert with fruit	3½ oz.	67	1.3	16.4	LC
Gelatin dessert with water	3½ oz.	59	1.5	14.1	LC
Gooseberries:					
Canned, heavy syrup	3½ oz.	117	.5	30.0	HC
Canned, water-packed	3½ oz.	26	0.5	6.6	LC
Raw	3½ oz.	39	0.8	9.7	LC
Granadilla, purple (passion-fruit)	3½ oz.	90	2.2	21.2	LC
Grapefruit, canned, water-packed	3½ oz.	70	.6	17.8	LC
Grapefruit, whole	3½ oz.	41	.5	10.6	LCR
Grapes	3½ oz.	69	1.3	15.7	LCR
Ground cherries (poha or cape-gooseberries)	3½ oz.	53	1.9	11.2	LCR
Guavas	3½ oz.	62	.8	15.0	LCR
Haws, scarlet	3½ oz.	87	2.0	20.8	LC
Highway Bartlett Pears	½ cup	80	0.5	20	LC
Highway Cling Peaches	½ cup	70	0.5	18	LC
Highway Unpeeled Apricots	½ cup	80	0.5	20	LC
Honeydew melon	3½ oz.	33	.8	7.7	LC
Hunt's Apricots	½ cup	103	0.77	26.4	HC
Hunt's Fruit Cocktail	½ cup	89	0.5	29.8	LC
Hunt's Peaches	½ cup	96	10.5	25.2	LC
Hunt's Pears	½ cup	90	0.25	23.2	LC

Fruits and Fruit Products

Food	Quantity	Calories	Protein Grams	Carbo-hydrates Grams	P/C/C Com-puter
Hunt's Snack Pack, Applesauce	1 can	94	0.3	24.4	LC
Hunt's Snack Pack, Mixed Fruit	1 can	96	0.51	24.8	LC
Hunt's Snack Pack, Peaches	1 can	106	0.57	27.7	HC
Jackfruit	3½ oz.	98	1.3	25.4	LC
Kumquats	3½ oz.	65	.9	17.1	LC
Lawry Snappy Canned Apples	3.5 oz.	383.6	1.8	88.3	HC
Lemons, peeled fruit	3½ oz.	27	1.1	8.2	LC
Lemons, fruit with peel	3½ oz.	20	1.2	10.7	LC
Libby's Fruit Cocktail, Heavy Syrup	½ cup	100	0.5	25	HC
Libby's Heavy Peaches in Syrup	½ cup	95	–	25	HC
Libby's Heavy Syrup Pears	½ cup	95	–	25	HC
Libby's Pumpkin	½ cup	40	1.5	10	LC
Limes, acid type, raw	3½ oz.	28	.7	9.5	LC
Loganberries:					
Canned, heavy syrup	3½ oz.	89	.6	22.2	HC
Canned, water-packed	3½ oz.	40	.7	9.4	LC
Fresh	3½ oz.	62	1.0	14.9	LC
Mamey (mammeeapple)	3½ oz.	51	.5	12.5	LC
Mangos	3½ oz.	66	.7	16.8	LC
Nectarines	3½ oz.	64	.6	17.1	LC
Oranges, all varieties, peeled fruit	3½ oz.	49	1.0	12.2	LC
Pantry Pride Applesauce	½ cup	100	0.5	23.5	HC
Pantry Pride Apricots	½ cup	100	0.5	23.5	HC
Pantry Price Cling Peaches in Heavy Syrup	½ cup	95	0.5	25	HC
Pantry Pride Fruit Cocktail	½ cup	100	0.5	25	HC
Pantry Pride Pears	½ cup	95	0	25	HC
Pantry Pride Pineapple	½ cup	70	0.5	17.5	LC
Pantry Pride Pumpkin	½ cup	40	1.5	10	LC
Pantry Pride Sweet Cherries	½ cup	70	0	17.5	LC
Papaws	3½ oz.	85	5.2	16.8	LC
Papayas	3½ oz.	39	.6	10.0	LC
Peaches:					
Canned, heavy-syrup	3½ oz.	78	.4	20.1	LC
Canned, water-packed	3½ oz.	31	.4	8.1	LC
Dried, nugget-type	3½ oz.	340	4.8	88.0	HCR
Fresh, raw	3½ oz.	38	.6	9.7	LC
Frozen, sliced, sweetened	3½ oz.	88	.4	22.6	HCR
Pears:					
Candied	3½ oz.	303	1.3	75.9	HC
Canned, heavy syrup	3½ oz.	76	.2	19.6	LC
Canned, juice-packed	3½ oz.	46	.3	11.8	LC
Canned, water-packed	3½ oz.	32	.2	8.3	LC

Food	Quantity	Calories	Protein Grams	Carbo-hydrates Grams	P/C/C Com-puter
Pears *(cont.)*:					
Dried	3½ oz.	268	3.1	67.3	HC
Raw with skin	3½ oz.	61	.7	15.3	LC
Persimmons	3½ oz.	127	.8	33.5	HC
Pineapple:					
Candied	3½ oz.	316	.8	80.0	HC
Canned, heavy-syrup	3½ oz.	90	.3	23.4	HCR
Canned, juice-packed	3½ oz.	58	.4	15.1	LC
Canned, water-packed	3½ oz.	39	.3	10.2	LC
Frozen chunks, sweetened	3½ oz.	85	.4	22.2	HCR
Raw	3½ oz.	52	0.4	13.7	LC
Pitanga (Surinam-cherry)	3½ oz.	51	.8	12.5	LC
Plantain (baking banana)	3½ oz.	119	1.1	31.2	HC
Plums:					
Fresh	3½ oz.	66	.5	17.8	LC
Purple, heavy-syrup	3½ oz.	83	.4	21.6	HCR
Purple, water-packed	3½ oz.	46	.4	11.9	LC
Pokeberry (poke) shoots, cooked, boiled, drained	3½ oz.	20	2.3	3.1	LC
Pomegranate pulp, raw	3½ oz.	63	.5	16.4	LC
Prunes:					
Dehydrated, cooked, with sugar	3½ oz.	180	1.2	47.1	HC
Dehydrated, cooked, without added sugar	3½ oz.	119	1.0	31.4	HC
Pumpkin, canned	3½ oz.	33	1.0	7.9	LC
Pumpkin, raw	3½ oz.	26	1.0	6.5	LC
Quinces	3½ oz.	57	.4	15.3	LC
Raisins, natural, unbleached, raw	3½ oz.	289	2.5	77.4	HC
Raisins cooked with sugar	3½ oz.	213	1.2	56.4	HC
Raspberries:					
Black	3½ oz.	73	1.5	15.7	LC
Canned black	3½ oz.	51	1.1	10.7	LC
Canned red	3½ oz.	35	.7	8.8	LC
Frozen red, sweetened	3½ oz.	98	.7	24.6	HC
Red	3½ oz.	57	1.2	13.6	LC
Rose Apples	3½ oz.	56	.6	14.2	LC
S & W Nutradiet Blue-Green Label Applesauce	3½ oz.	48	0.28	11.46	LC
S & W Nutradiet Blue-Green Label Apricots	3½ oz.	38	0.55	8.69	LC
S & W Nutradiet Blue-Green Label Fruit Cocktail	3½ oz.	35	0.48	10.65	LC

Fruits and Fruit Products

Food	Quantity	Calories	Protein Grams	Carbo-hydrates Grams	P/C/C Computer
S & W Nutradiet Blue-Green Label Grapefruit Sections	3½ oz.	36	0.73	1.47	LC
S & W Nutradiet Blue-Green Label Mandarin Oranges	3½ oz.	27	0.46	6.14	LC
S & W Nutradiet Blue-Green Label Peach Halves	3½ oz.	28	0.53	0.27	LC
S & W Nutradiet Blue-Green Label Peach Slices	3½ oz.	24	0.53	5.11	LC
S & W Nutradiet Blue-Green Label Pear Halves	3½ oz.	30	0.26	7.15	LC
S & W Nutradiet Blue-Green Label Pear Quarters	3½ oz.	27	0.22	6.46	LC
S & W Nutradiet Blue-Green Label Pineapple Chunks	3½ oz.	49	0.03	11.07	LC
S & W Nutradiet Blue-Green Label Pineapple Tidbits	3½ oz.	56	0.03	13.05	LC
S & W Nutradiet Blue-Green Label Royal Anne Cherries	3½ oz.	47	0.88	10.65	LC
S & W Nutradiet Blue-Green Label Salad Fruits	3½ oz.	38	0.46	8.67	LC
S & W Nutradiet Blue-Green Label Whole Figs	3½ oz.	52	0.5	12.24	LC
S & W Nutradiet Red Label Applesauce	3½ oz.	48	0.21	11.45	LC
S & W Nutradiet Red Label Apricot Halves	3½ oz.	40	0.59	9.14	LC
S & W Nutradiet Red Label Blackberries	3½ oz.	36	0.9	7.88	LC
S & W Nutradiet Red Label Boysenberries	3½ oz.	32	0.8	7.65	LC
S & W Nutradiet Red Label Cling Peach Halves	3½ oz.	25	0.42	5.74	LC
S & W Nutradiet Red Label Cling Peach Slices	3½ oz.	25	0.45	5.6	LC
S & W Nutradiet Red Label Dark Sweet Cherries	3½ oz.	53	0.55	12.49	LC
S & W Nutradiet Red Label Figs	3½ oz.	49	0.53	11.51	LC
S & W Nutradiet Red Label Freestone Peach Halves	3½ oz.	26	0.47	5.92	LC
S & W Nutradiet Red Label Freestone Peach Slices	3½ oz.	24	0.5	5.49	LC
S & W Nutradiet Red Label Fruit Cocktail	3½ oz.	36	0.56	8.18	LC
S & W Nutradiet Red Label Mandarin Oranges	3½ oz.	27	0.46	6.14	LC
S & W Nutradiet Red Label Pear Halves	3½ oz.	28	0.3	6.53	LC

Food	Quantity	Calories	Protein Grams	Carbo-hydrates Grams	P/C/C Com-puter
S & W Nutradiet Red Label Pear Quarters	3½ oz.	26	0.2	6.15	LC
S & W Nutradiet Red Label Pineapple Chunks	3½ oz.	69	0.41	16.67	LC
S & W Nutradiet Red Label Pineapple Tidbits	3½ oz.	69	0.43	16.55	LC
S & W Nutradiet Red Label Purple Plums	3½ oz.	50	0.46	11.88	LC
S & W Nutradiet Red Label Royal Anne Cherries	3½ oz.	49	0.81	11.29	LC
S & W Nutradiet Red Label Salad Fruits	3½ oz.	35	0.47	7.94	LC
S & W Nutradiet Red Label Strawberries	3½ oz.	20	0.55	4.33	LC
S & W Nutradiet Red Label Whole Apricots	3½ oz.	31	0.6	6.88	LC
Sapodilla	3½ oz.	89	.5	21.8	LC
Sapotes (marmalade plums)	3½ oz.	125	1.8	31.6	HC
Smucker's Dutch Girl Apple Butter	1 tsp.	12	0	3	LC
Smucker's Jellies	1 tsp.	18	0	4	LC
Smucker's Orange Marmalade	1 tsp.	18	0	4	LC
Smucker's Peach Butter	1 tsp.	16	0	4	LC
Smucker's Preserves	1 tsp.	18	0	4	LC
Smucker's Spiced and Cider Apple Butter	1 tsp.	14	0	4	LC
Soursop	3½ oz.	65	1.0	16.3	LC
Stokely-Van Camp Applesauce	½ cup	109	0.2	28.5	HC
Stokely-Van Camp Apricots	½ cup	103	0.7	26	HC
Stokely-Van Camp Figs	½ cup	100	0.6	26.1	HC
Stokely-Van Camp Fruit Cocktail	½ cup	87	0.5	22.6	LC
Stokely-Van Camp Fruits for Salad	½ cup	90	0.4	22.3	LC
Stokely-Van Camp Grapefruit Sections	½ cup	80	0.7	20.4	LC
Stokely-Van Camp Peaches	½ cup	89	0.5	23.1	LC
Stokely-Van Camp Pears	½ cup	87	0.2	22.5	LC
Stokely-Van Camp Pineapple	½ cup	84	0.4	22.3	LC
Stokely-Van Camp Pitted Sour Cherries	½ cup	49	0.9	17.2	LC
Stokely-Van Camp Pumpkin	½ cup	38	1.1	9	LC
Stokely-Van Camp Sweet Cherries	½ cup	97	1.0	24.6	LC
Strawberries:					
Canned	3½ oz.	22	.4	5.6	LC
Fresh	3½ oz.	37	.7	8.4	LC
Frozen	3½ oz.	92	.4	23.5	LC
Sugar apples (sweetsop) fresh	3½ oz.	94	1.8	23.7	LC

Fruits and Fruit Products

Food	Quantity	Calories	Protein Grams	Carbo-hydrates Grams	P/C/C Computer
Tamarinds, fresh	3½ oz.	239	2.8	62.5	HC
Tangerines, fresh	3½ oz.	46	.8	11.6	LC
Tillie Lewis Tasti-Diet Applesauce	½ cup	58	0	15	LC
Tillie Lewis Tasti-Diet Apricots	½ cup	60	0	15	LC
Tillie Lewis Tasti-Diet Bartlett Pears	½ cup	50	0	13	LC
Tillie Lewis Tasti-Diet Elberta Peaches	½ cup	45	0	12	LC
Tillie Lewis Tasti-Diet Fruit Cocktail	½ cup	50	0	13	LC
Tillie Lewis Tasti-Diet Grapefruit	½ cup	42	0	10	LC
Tillie Lewis Tasti-Diet Kadota Figs	½ cup	60	0	14	LC
Tillie Lewis Tasti-Diet Light Sweet Cherries	½ cup	60	0	14	LC
Tillie Lewis Tasti-Diet Mandarin Oranges	½ cup	31	0	7	LC
Tillie Lewis Tasti-Diet Pineapple	½ cup	77	0	20	LC
Tillie Lewis Tasti-Diet Purple Plums	½ cup	54	0	16	LC
Tillie Lewis Tasti-Diet Y.C. Peaches	½ cup	45	0	11	LC
Town House Applesauce, Gravenstein	½ cup	85	0	23	LC
Town House Bartlett Pears	½ cup	95	0.5	24.5	HCR
Town House Cling Peaches	½ cup	95	0.5	25	HCR
Town House Freestone Peaches	½ cup	130	0.5	34	HCR
Town House Fruit Cocktail	½ cup	85	0.5	23	LC
Town House Pineapple in Syrup	½ cup	100.5	0.7	26.8	HC
Town House Unpeeled Apricots	½ cup	110	0.5	28	HC
Watermelon, fresh	3½ oz.	26	.5	6.4	LC
Wax gourd (Chinese melon) fresh	3½ oz.	13	.4	3.0	LC

Vegetables
and Vegetable Products

Food	Quantity	Calories	Protein Grams	Carbo-hydrates Grams	P/C/C Computer
Artichokes, globe or French	3½ oz.	22	2.8	9.9	LC
Asparagus	3½ oz.	20	2.2	3.6	LC
Bamboo shoots, raw	3½ oz.	27	2.6	5.2	LC
Beet greens	3½ oz.	18	1.7	3.3	LC
Beets, boiled, drained	3½ oz.	32	1.1	7.2	LC
Beets, canned, drained	3½ oz.	37	1.0	8.8	LC
Birds Eye Baby Butter Beans	3½ oz.	130	8.9	2.5	HC
Birds Eye Black Eye Peas	½ cup	92	6.4	15.7	LC
Birds Eye Candied Sweet Potatoes	3½ oz.	188	1.2	46.6	HCR
Birds Eye Chopped Onions	¼ cup	11	0.4	2.5	LC
Birds Eye Collard Greens	3½ oz.	30.7	3.1	4.8	LC
Birds Eye Combination Asparagus Spears w/Hollandaise Sauce	3½ oz.	103	3.4	3.4	HC
Birds Eye Combination Broccoli Spears w/Hollandaise Sauce	3½ oz.	106	3.4	3.4	HC
Birds Eye Combination Carrots w/Brown Sugar	½ cup	87	0.8	16.8	LC
Birds Eye Combination Corn and Peas w/Tomatoes	3½ oz.	71	3.2	17	LC
Birds Eye Combination Mixed Vegetables w/Onion Sauce	3½ oz.	124	3.1	14.4	HC
Birds Eye Combination Peas w/Cream Sauce	3½ oz.	162	5.6	17.6	HC
Birds Eye Combination Peas and Celery	½ cup	55	3.6	9.8	LC
Birds Eye Combination Peas and Pearl Onions	3½ oz.	71	4.45	13.0	LC
Birds Eye Combination Peas and Potatoes w/Cream Sauce	3½ oz.	139	3.7	15.8	HC
Birds Eye Combination Peas and Rice w/Mushrooms	½ cup	113	3.4	21.9	HC
Birds Eye Combination Peas w/Mushrooms	3½ oz.	85.5	6.1	15.2	LC
Birds Eye Combination Sliced Beets w/Orange Glaze	3½ oz.	82	1.7	22.6	LC
Birds Eye Combination Small Onions w/Cream Sauce	3½ oz.	140	2.4	13.5	HC
Birds Eye Combination Sweet Potatoes w/Brown Sugar, Pineapple	3½ oz.	168	1.6	35.6	HC
Birds Eye Combination Vegetable Jubilee	½ cup	138	3	18.9	HC
Birds Eye Corn on Cob	1 ear	98	3.6	21.7	HCR

Food	Quantity	Calories	Protein Grams	Carbo-hydrates Grams	P/C/C Computer
Birds Eye Cut Okra	½ cup	36	2.5	7.6	LC
Birds Eye Deluxe Artichoke Hearts	3½ oz.	25.5	1.4	5.6	LC
Birds Eye Deluxe Broccoli Spears	3½ oz.	28	3.3	3.8	LC
Birds Eye Deluxe Brussels Sprouts	½ cup	34	3.1	5.7	LC
Birds Eye Deluxe Creamed Spinach	3½ oz.	71	2.7	6.3	LC
Birds Eye Deluxe Sweet White Corn	½ cup	77	2.9	17.9	LC
Birds Eye Deluxe Tiny Lima Beans	½ cup	83	5.4	16.2	LC
Birds Eye Deluxe Tiny Tender Peas	½ cup	70	5.1	12	LC
Birds Eye Deluxe Whole Green Beans	3½ oz.	27	1.6	6.0	LC
Birds Eye 5-Minute Asparagus Spears	3½ oz.	23	3.2	3.8	LC
Birds Eye 5-Minute Baby Lima Beans	½ cup	111	6.9	21	HC
Birds Eye 5-Minute Broccoli Spears	3½ oz.	28	3.2	3.8	LC
Birds Eye 5-Minute Cauliflower	3½ oz.	22	2.1	3.4	LC
Birds Eye 5-Minute Chopped Broccoli	3½ oz.	28.6	3.1	3.8	LC
Birds Eye 5-Minute Chopped Spinach	3½ oz.	24	3	2.9	LC
Birds Eye 5-Minute Cut Asparagus	½ cup	20	3.0	3.3	LC
Birds Eye 5-Minute Cut Green Beans	3½ oz.	25.5	1.6	5.9	LC
Birds Eye 5-Minute Cut Wax Beans	3½ oz.	28	1.6	6.3	LC
Birds Eye 5-Minute Fordhook Lima Beans	3½ oz.	100	6.0	19.0	LC
Birds Eye 5-Minute French Style Green Beans	3½ oz.	25.5	1.6	5.9	LC
Birds Eye 5-Minute French Green Beans w/Almonds	½ cup	52	2.1	6	LC
Birds Eye Deluxe French Green Beans w/Mushrooms	3½ oz.	30	1.6	6.4	LC
Birds Eye 5-Minute Italian Green Beans	3½ oz.	27	2.1	4.8	LC
Birds Eye 5-Minute Leaf Spinach	3½ oz.	24	3	3.3	LC

Food	Quantity	Calories	Protein Grams	Carbo-hydrates Grams	P/C/C Com-puter
Birds Eye 5-Minute Mixed Vegetables	½ cup	50	3.0	124	HCR
Birds Eye 5-Minute Peas and Carrots	½ cup	55	3.0	8.7	LC
Birds Eye 5-Minute Succotash	½ cup	87	4.1	19.4	LC
Birds Eye 5-Minute Sweet Corn	½ cup	77	2.8	17.8	LC
Birds Eye 5-Minute Zucchini	3½ oz.	21	1	2.4	LC
Birds Eye French Fried Onion Rings	3½ oz.	290.5	4.55	29.9	HC
Birds Eye International, Bavarian Style	3½ oz.	143	3.7	12.5	HC
Birds Eye International, Chinese Style	3½ oz.	73	2.65	6.4	LC
Birds Eye International, Danish Style	3½ oz.	97.5	1.8	8.1	HC
Birds Eye International, Hawaiian Style	3½ oz.	97.5	0.95	13.4	HC
Birds Eye International, Italian Style	3½ oz.	104	2.8	8.4	HC
Birds Eye International, Japanese Style	3½ oz.	111	1.9	6.1	HC
Birds Eye International, Mexican Style	3½ oz.	153	4.7	18.0	HC
Birds Eye International, Parisian Style	3½ oz.	95	1.4	7.8	HC
Birds Eye International, Spanish Style	3½ oz.	90	1.2	7.5	HC
Birds Eye Kale	½ cup	29	3.0	4.2	LC
Birds Eye Mustard Greens	½ cup	19	2.2	2.2	LC
Birds Eye Small Whole Onions	½ cup	51	1.3	12	LC
Birds Eye Squash	3½ oz.	38	1.2	8.0	LC
Birds Eye Sweet Peas	½ cup	70	5.0	12.2	LC
Birds Eye Turnip Greens	½ cup	22	2.4	2.7	LC
Birds Eye Whole Okra	½ cup	27	1.9	5.7	LC
Broccoli, cooked, drained	3½ oz.	26	3.1	4.5	LC
Broccoli, frozen, cooked, drained	3½ oz.	26	2.9	4.6	LC
Brussels sprouts, cooked, drained	3½ oz.	36	4.2	6.4	LC
Brussels sprouts, frozen, cooked, drained	3½ oz.	33	3.2	6.5	LC
Cabbage, boiled, drained	3½ oz.	18	1.0	4.0	LC
Campbell's Tomato Wedges	½ cup	28.5	0.95	6.3	LC
Carrots:					
Boiled, drained	3½ oz.	31	.9	7.1	LC
Canned, drained	3½ oz.	30	.8	6.7	LC
Raw	3½ oz.	42	1.1	9.7	LC

Food	Quantity	Calories	Protein Grams	Carbo-hydrates Grams	P/C/C Computer
Cauliflower:					
Boiled, drained	3½ oz.	22	2.3	4.1	LC
Frozen, boiled, drained	3½ oz.	18	1.9	3.3	LC
Raw	3½ oz.	27	2.7	5.2	LC
Celeriac root	3½ oz.	40	1.8	8.5	LC
Celery, boiled, drained	3½ oz.	14	.8	3.1	LC
Celery, raw	3½ oz.	17	.9	3.9	LC
Chard, Swiss, boiled, drained	3½ oz.	18	1.8	3.3	LC
Chard, Swiss, raw	3½ oz.	25	2.4	4.6	LC
Chervil	3½ oz.	57	3.4	11.5	LC
Chicory (endive)	3½ oz.	15	1.0	3.2	LC
Chicory greens, raw	3½ oz.	20	1.8	3.8	LC
Chinese cabbage (also called celery cabbage or petsai), raw	3½ oz.	16	1.6	2.9	LC
Chives, raw	3½ oz.	28	1.8	5.8	LC
Coconut cream	3½ oz.	334	4.4	8.3	HC
Coconut meat, fresh	3½ oz.	346	3.5	9.4	HC
Coconut meat, dried	3½ oz.	662	7.2	23.0	HC
Coleslaw with commercial French dressing	3½ oz.	95	1.2	7.6	HC
Coleslaw with home-made French dressing	3½ oz.	129	1.1	5.1	HC
Coleslaw with mayonnaise	3½ oz.	144	1.3	4.8	HC
Coleslaw with mayonnaise-type salad dressing	3½ oz.	99	1.2	7.1	HC
Corn:					
Canned, cream-style	3½ oz.	82	2.1	20.0	HCR
Field, whole-grain	3½ oz.	348	8.9	72.2	HCR
Frozen, boiled, drained	3½ oz.	79	3.0	18.8	LC
Kernels, frozen	3½ oz.	94	3.5	21.6	LC
Sweet, boiled, drained	3½ oz.	83	3.2	18.8	LC
Corn salad	3½ oz.	21	2.0	3.6	LC
Cress, garden, cooked, boiled, drained	3½ oz.	23	1.9	3.8	LC
Cucumbers	3½ oz.	15	.9	3.4	LC
Collards, boiled, drained	3½ oz.	31	3.4	4.8	LC
Collards, frozen, boiled, drained	3½ oz.	29	2.7	4.9	LC
Dandelion greens, boiled, drained	3½ oz.	33	2.0	6.4	LC
Del Monte Canned Asparagus	½ cup	17	1.6	3	LC
Del Monte Canned Beets	½ cup	26.5	0.7	6.05	LC
Del Monte Carrots	½ cup	23	0.65	4.95	LC

Food	Quantity	Calories	Protein Grams	Carbo-hydrates Grams	P/C/C Computer
Del Monte Corn 'n Peppers	½ cup	87.5	2.65	19.3	LC
Del Monte Cream-Style Corn	3½ oz.	111.5	2.75	25	HC
Del Monte Family Style Corn	½ cup	78.5	2.25	16.4	LC
Del Monte Hot Chili Peppers	½ cup	21.5	0.65	4.1	LC
Del Monte Mild Sweet Peppers	½ cup	21	0.55	4.55	LC
Del Monte Mixed Vegetables	½ cup	41	2.65	7.1	LC
Del Monte Peas	½ cup	43	2.9	7.3	LC
Del Monte Peas and Carrots	½ cup	37.5	2.1	6.75	LC
Del Monte Pickled Beets	½ cup	62.5	0.6	14.8	LC
Del Monte Sauerkraut	½ cup	18	0.9	3.75	LC
Del Monte Seasoned Peas	½ cup	52.5	4.1	8.7	LC
Del Monte Spinach	½ cup	24.5	3.35	3	LC
Del Monte Stewed Tomatoes	½ cup	30	1.05	5	LC
Del Monte Tomatoes	½ cup	30	1.05	5	LC
Del Monte Vacuum-Packed Corn	½ cup	91	2.65	19.6	HCR
Del Monte Zucchini in Tomato Sauce	½ cup	25	1.3	5.55	LC
Dock (curly, broadleaf, or sheep sorrel)	3½ oz.	19	1.6	3.9	LC
Durkee French Fried Onions	3½ oz.	618	0.6	44.5	HC
Durkee Potato Sticks	1¾ oz.	282	2.6	24.8	HC
Eggplant, boiled, drained	3½ oz.	19	1.0	4.1	LC
Endive	3½ oz.	20	1.7	4.1	LC
Fennel	3½ oz.	28	2.8	5.1	LC
French's Au Gratin Potato Mix	½ cup	95	3.7	15	LC
French's Country Style Mashed Potato Mix	½ cup	137	2.7	16.5	HC
French's Idaho Mashed Potato Mix	½ cup	114	2.4	16	HC
French's Pancakes Potato Mix	3 cakes	90	3.1	15.5	LC
French's Scalloped Potato Mix	½ cup	109	3.3	20	HC
Garlic, cloves	3½ oz.	137	6.2	30.8	HCR
Green Giant April Showers Early June Peas	3½ oz.	50	3.0	9.0	LC
Green Giant Boil-in-Bag Broccoli in Cheese Sauce	3½ oz.	60	4.0	6.0	LC
Green Giant Boil-in-Bag Broccoli Spears in Butter Sauce	3½ oz.	45	2.2	4.0	LC
Green Giant Boil-in-Bag Brussel Sprouts in Butter Sauce	3½ oz.	60	3.0	5.0	LC
Green Giant Broccoli and Noodle Casserole	3½ oz.	100	3.7	9.0	HC
Green Giant Boil-in-Bag Carrot Nuggets in Butter Sauce	3½ oz.	50	0.6	6.5	LC
Green Giant Boil-in-Bag Cauliflower in Butter Sauce	3½ oz.	40	1.4	3.0	LC

Vegetables and Vegetable Products

Food	Quantity	Calories	Protein Grams	Carbo-hydrates Grams	P/C/C Com-puter
Green Giant Boil-in-Bag Cauliflower in Cheese Sauce	3½ oz.	60	3.7	6.0	LC
Green Giant Boil-in-Bag Florida Style	3½ oz.	50	1.0	4.5	LC
Green Giant Boil-in-Bag Frozen Asparagus	3½ oz.	50	2.2	3.0	LC
Green Giant Boil-in-Bag Golden Corn, Cream-Style	3½ oz.	75	2.0	16.0	LC
Green Giant Boil-in-Bag Green Beans	3½ oz.	40	1.0	3.5	LC
Green Giant Boil-in-Bag Green Beans in Mushroom Sauce	3½ oz.	40	2.0	6.0	LC
Green Giant Boil-in-Bag Green Beans w/Onions and Bacon	3½ oz.	40	1.5	4.5	LC
Green Giant Boil-in-Bag Italian Green Beans in Tomato Sauce	3½ oz.	45	2.0	7.0	LC
Green Giant Boil-in-Bag LeSueur Peas in Butter Sauce	3½ oz.	80	3.5	10	LC
Green Giant Boil-in-Bag Lima Beans in Butter Sauce	3½ oz.	120	5.0	17	HC
Green Giant Boil-in-Bag Mexicorn	3½ oz.	90	2.2	15.0	HC
Green Giant Boil-in-Bag Mixed Vegetables in Butter Sauce	3½ oz.	65	2.3	10.0	LC
Green Giant Boil-in-Bag Mushrooms in Butter	3½ oz.	55	1.5	3.0	LC
Green Giant Boil-in-Bag New Orleans Style Vegetables	3½ oz.	80	3.0	8.0	LC
Green Giant Boil-in-Bag Northwest Style Vegetables	3½ oz.	85	2.8	10.0	LC
Green Giant Boil-in-Bag Onions in Cream Sauce	3½ oz.	45	2.2	7.0	LC
Green Giant Boil-in-Bag Pennsylvania Dutch Style Vegetables	3½ oz.	100	2.2	15.0	LC
Green Giant Boil-in-Bag San Francisco Style Vegetables	3½ oz.	35	1.5	5.0	LC
Green Giant Boil-in-Bag Sliced Carrots in Honey Glaze	3½ oz.	75	0.7	14.0	LC
Green Giant Boil-in-Bag Spinach in Butter Sauce	3½ oz.	45	2.5	2.5	LC

Food	Quantity	Calories	Protein Grams	Carbo-hydrates Grams	P/C/C Com-puter
Green Giant Boil-in-Bag Spinach in Cream Sauce	3½ oz.	60	2.5	5.0	LC
Green Giant Boil-in-Bag Sweet Peas in Butter Sauce	3½ oz.	80	4.0	10	LC
Green Giant Boil-in-Bag Sweet Peas in Cream Sauce	3½ oz.	65	4.0	10	LC
Green Giant Boil-in-Bag Sweet Peas w/Carrots in Cream Sauce	3½ oz.	55	3.5	9	LC
Green Giant Boil-in-Bag Sweet Peas w/Onions in Butter Sauce	3½ oz.	75	3.5	10	LC
Green Giant Boil-in-Bag White Whole Corn Kernel in Butter Sauce	3½ oz.	90	2.3	16.0	LC
Green Giant Brussels Sprouts Au Gratin Casserole	3½ oz.	70	4.4	8.0	LC
Green Giant Canned Asparagus	3½ oz.	15	1.7	1.7	LC
Green Giant Corn Kernel in Liquid	3½ oz.	65	1.8	13	LC
Green Giant Corn Niblets	3½ oz.	85	2.4	17.0	LC
Green Giant Cream-Style Corn	3½ oz.	90	1.7	19	LC
Green Giant Deviled Spinach Casserole	3½ oz.	70	4.5	5	LC
Green Giant Hungarian Cauliflower Casserole	3½ oz.	70	2.5	7.0	LCR
Green Giant June Peas w/Onions	3½ oz.	45	3.0	10.0	LC
Green Giant LeSueur Canned Asparagus Spears	3½ oz.	15	1.7	1.7	LC
Green Giant LeSueur Golden Corn Shoe Peg	3½ oz.	65	1.8	13.0	LC
Green Giant LeSueur Small Carrots	½ cup	20	0.4	4.5	LC
Green Giant LeSueur Small Early June Peas	3½ oz.	45	2.7	8.0	LC
Green Giant Mexicorn	3½ oz.	85	2.4	17.0	LC
Green Giant Mushrooms	3½ oz.	25	1.5	3.0	LC
Green Giant Niblets Ears	3½ oz.	110	3.0	25.6	HC
Green Giant Scalloped Corn Casserole	3½ oz.	140	4.5	16.0	HC
Green Giant Sweet Peas	3½ oz.	45	3.0	8.0	LC
Green Giant White Corn, Whole Kernel	3½ oz.	85	2.3	18	LC
Holloway House Potatoes w/Cheese	3½ oz.	160	2.5	19	HC
Holloway House Potatoes w/Sour Cream	3½ oz.	140	2.5	19	HC

Vegetables and Vegetable Products

Food	Quantity	Calories	Protein Grams	Carbo-hydrates Grams	P/C/C Com-puter
Holloway House Baked Potatoes w/Cheese	3½ oz.	170	3.0	22	HC
Holloway House Baked Potatoes w/Sour Cream	3½ oz.	170	3.0	22	HC
Hunt's Barbeque Skillet Mix w/o Meat	3½ oz.	150	5.17	25.6	HC
Hunt's Peeled Whole Tomatoes	½ cup	20	1.0	4.35	LC
Hunt's Snack Pack Potato Salad	5 oz.	225	1.98	26	HC
Hunt's Stewed Tomatoes	3½ oz.	30	1.06	6.9	LC
Jerusalem artichoke	3½ oz.	50	2.3	16.7	LC
Kale, cooked, boiled, drained	3½ oz.	39	4.5	6.1	LC
Kale, frozen, cooked, boiled, drained	3½ oz.	31	3.0	5.4	LC
Kohlrabi, cooked, boiled, drained	3½ oz.	24	1.7	5.3	LC
La Choy Meatless Chow Mein	1 cup	49	3.1	5.4	LC
La Choy Mushroom Chow Mein Bi-Pack	1 cup	81	3.3	7.7	LC
La Choy Vegetable Chow Mein	3½ oz.	28.78	1.02	7.03	LC
Lamb's-quarters, cooked, boiled, drained	3½ oz.	32	3.2	5.0	LC
Leeks, bulb and lower leaf portions	3½ oz.	52	2.2	11.2	LC
Lettuce, raw	3½ oz.	14	1.2	2.5	LC
Libby's Canned Beets	½ cup	32.5	1	7.5	LC
Libby's Cream-Style Corn	½ cup	90	2	22.5	LC
Libby's Harvard Beets	½ cup	75	1	17.5	LC
Libby's Pickled Beets	½ cup	80	1	20	LC
Libby's Sauerkraut	½ cup	22.5	1	5	LC
Libby's Whole Corn	½ cup	85	2.5	20	LC
Mrs. Paul's Candied Sweet Potatoes	3½ oz.	162.4	0.86	39.67	HC
Mrs. Paul's Corn Fritters	3½ oz.	246.75	4.75	36.24	HC
Mrs. Paul's Eggplant Parmesan	3½ oz.	201.25	5.55	15.17	HC
Mrs. Paul's Fried Eggplant	3½ oz.	165	3.52	25.34	HC
Mrs. Paul's Fried Onion Rings	3½ oz.	220.15	4.92	19.77	HC
Mrs. Paul's Fried, Breaded Onion Rings	3½ oz.	292.6	4.74	29.15	HC
Mrs. Paul's Fried Zucchini	3½ oz.	220	4.4	24.5	HC
Mrs. Paul's Zucchini Parmesan	3½ oz.	75.6	5.4	4.7	LC
Mushrooms, raw	3½ oz.	28	2.7	4.4	LC
Mushrooms, canned	3½ oz.	17	1.9	2.4	LC
Mustard greens, cooked, boiled, drained	3½ oz.	23	2.2	4.0	LC
Mustard spinach (tender, green) cooked, boiled, drained	3½ oz.	16	1.7	2.8	LC

Food	Quantity	Calories	Protein Grams	Carbo-hydrates Grams	P/C/C Computer
Okra, cooked, boiled, drained	3½ oz.	29	2.0	6.0	LC
Okra, frozen, cooked, boiled, drained	3½ oz.	38	2.2	8.8	LC
Olives, pickled, canned, bottled, green	3½ oz.	116	1.4	1.3	HC
Onions:					
Cooked, boiled, drained	3½ oz.	29	1.2	6.5	LC
Dehydrated, flaked	3½ oz.	350	8.7	82.1	HC
Young, green (bunching varieties)	3½ oz.	36	1.5	8.2	LC
Oregon Freeze Dry Potato Chowder	3½ oz.	390	14.5	66	HC
Pantry Pride Canned Beets	½ cup	35	1	7.5	LC
Pantry Pride Cream-Style Corn	½ cup	100	2.45	21.5	HC
Pantry Pride Peas	½ cup	72.5	4	13	LC
Pantry Pride Sweet Potatoes	½ cup	115	1	28	HC
Pantry Pride Tomatoes w/Juice	½ cup	25	1	59.5	LC
Pantry Pride Whole Corn	½ cup	85	2.45	17.5	LC
Parsley	3½ oz.	44	3.6	8.5	LC
Parsnips, cooked, boiled, drained	3½ oz.	66	1.5	14.9	LC
Parsnips, raw	3½ oz.	76	1.7	17.5	LC
Peas:					
Canned, drained solids	3½ oz.	88	4.7	16.8	LC
Canned, sweet, drained solids	3½ oz.	80	4.6	15.0	LC
Cooked, boiled, drained	3½ oz.	43	2.9	9.5	LC
Frozen, drained solids	3½ oz.	68	5.1	11.8	LC
Green, immature, cooked, boiled, drained	3½ oz.	71	5.4	12.1	LC
Peas and carrots, frozen, cooked, boiled, drained	3½ oz.	53	3.2	10.1	LC
Peppers, hot, chili	3½ oz.	37	1.3	9.1	LC
Peppers, sweet, green	3½ oz.	18	1.0	3.8	LC
Pickles:					
Cucumber	3½ oz.	73	.9	17.9	LC
Dill	3½ oz.	11	.7	2.2	LC
Sour	3½ oz.	10	.5	2.0	LC
Sweet	3½ oz.	146	.7	36.5	HC
Pillsbury Hungry Jack Au Gratin Potato Mix	½ cup	170	5	21	HC

Vegetables and Vegetable Products

Food	Quantity	Calories	Protein Grams	Carbo- hydrates Grams	P/C/C Com- puter
Pillsbury Hungry Jack Complete Mashed Potato Mix	½ cup	150	3	18	HC
Pillsbury Hungry Jack Hash Brown Potato Mix	½ cup	140	2	25	HC
Pillsbury Hungry Jack Mashed Potato Mix	½ cup	170	3	18	HC
Pillsbury Hungry Jack Potato Mix	½ cup	140	3	21	HC
Pimientos, canned, solids and liquids	3½ oz.	27	.9	5.8	LC
Potato salad, home recipe, with cooked salad dressing	3½ oz.	99	2.7	16.3	LC
Potatoes:					
Baked in skin	3½ oz.	93	2.6	21.1	HCR
Boiled in skin	3½ oz.	76	2.1	17.1	LC
Boiled, pared before cooking	3½ oz.	65	1.9	14.5	LC
Canned, solids and liquid	3½ oz.	44	1.1	9.8	LC
Dehydrated mashed without milk	3½ oz.	364	7.2	84.0	HC
Dehydrated mashed with milk and table fat added	3½ oz.	93	1.9	14.5	HC
French-fried	3½ oz.	274	4.3	36.0	HC
Fried from raw	3½ oz.	268	4.0	32.6	HC
Frozen, cooked, hash-browned	3½ oz.	224	2.0	29.0	HC
Frozen, French-fried	3½ oz.	220	3.6	33.7	HC
Hash-browned after holding overnight	3½ oz.	229	3.1	29.1	HC
Mashed, milk added	3½ oz.	65	2.1	13.0	LC
Mashed, milk and table fat added	3½ oz.	94	2.1	12.3	HC
Scalloped and au gratin with cheese	3½ oz.	145	5.3	13.6	HC
Scalloped and au gratin without cheese	3½ oz.	104	3.0	14.7	HC
Purslane leaves, cooked, boiled, drained	3½ oz.	15	1.2	2.8	LC
Radishes	3½ oz.	17	1.0	3.6	LC

Food	Quantity	Calories	Protein Grams	Carbo-hydrates Grams	P/C/C Computer
Red cabbage, raw	3½ oz.	31	2.0	6.9	LC
Rhubarb, cooked with sugar	3½ oz.	141	.5	36.0	HC
Rhubarb, frozen, cooked, added sugar	3½ oz.	143	.5	32.2	HC
Rutabagas, cooked, boiled, drained	3½ oz.	35	.9	8.2	LC
S & W Nutradiet Blue-Green Label Asparagus	3½ oz.	16	1.91	1.96	LC
S & W Nutradiet Blue-Green Label Beans	3½ oz.	16	0.98	2.81	LC
S & W Nutradiet Blue-Green Label Beets	3½ oz.	28	0.83	5.93	LC
S & W Nutradiet Blue-Green Label Carrots	3½ oz.	22	0.55	4.64	LC
S & W Nutradiet Blue-Green Label Cream-Style Corn	3½ oz.	84	2.56	17.33	LC
S & W Nutradiet Blue-Green Label Peas and Carrots	3½ oz.	32	1.94	5.74	LC
S & W Nutradiet Blue-Green Label Sweet Peas	3½ oz.	35	2.64	5.96	LC
S & W Nutradiet Blue-Green Label Whole Kernel Corn	3½ oz.	52	1.63	10.43	LC
S & W Nutradiet Blue-Green Label Whole Tomatoes	3½ oz.	21	0.98	4.1	LC
Salsify, cooked, boiled, drained	3½ oz.	40	2.6	15.1	LC
Sauerkraut, canned	3½ oz.	18	1.0	4.0	LC
Savoy cabbage, raw	3½ oz.	24	2.4	4.6	LC
Shady Oak Mushrooms	3½ oz.	16.6	1.75	1.75	LC
Shallot bulbs	3½ oz.	72	2.5	16.8	LC
Spinach:					
Canned	3½ oz.	19	2.0	3.0	LC
Fresh, cooked, boiled, drained	3½ oz.	23	3.0	3.6	LC
Frozen	3½ oz.	23	3.0	3.7	LC
Squash:					
Acorn	3½ oz.	55	1.9	14.0	LC
Butternut	3½ oz.	68	1.8	17.5	LC
Crookneck and straightneck, yellow	3½ oz.	15	1.0	3.1	LC
Hubbard	3½ oz.	50	1.8	11.7	LC
Scallop varieties	3½ oz.	16	.7	3.8	LC
Summer	3½ oz.	14	.9	3.1	LC
Winter	3½ oz.	63	1.8	15.4	LC
Zucchini	3½ oz.	12	1.0	2.5	LC

Vegetables and Vegetable Products

Food	Quantity	Calories	Protein Grams	Carbo-hydrates Grams	P/C/C Com-puter
Squash, frozen:					
Summer, crookneck	3½ oz.	21	1.4	4.7	LC
Winter	3½ oz.	38	1.2	9.2	LC
Stokely-Van Camp Asparagus	½ cup	20	2.1	3.3	LC
Stokely-Van Camp Beets	½ cup	39	1.0	9.0	LC
Stokely-Van Camp Carrots	½ cup	32	0.7	7.4	LC
Stokely-Van Camp Cream-Style Corn	½ cup	94	2.4	23	HC
Stokely-Van Camp Early June Peas	½ cup	76	4	14.3	LC
Stokely-Van Camp Mixed Vegetables	½ cup	71	3.6	15	LC
Stokely-Van Camp Sauerkraut	½ cup	20	1.1	4.4	LC
Stokely-Van Camp Spinach	½ cup	21	2.2	3.3	LC
Stokely-Van Camp Sweet Peas	½ cup	65	3.9	11.9	LC
Stokely-Van Camp Turnip Greens	½ cup	20	1.6	3.5	LC
Stokely-Van Camp Whole Corn	½ cup	74	2.2	18	LC
Stokely-Van Camp Whole Tomatoes	½ cup	23	1.1	4.9	LC
Succotash (corn and lima beans) frozen	3½ oz.	93	4.2	20.5	HCR
Swamp cabbage, fresh	3½ oz.	29	3.0	5.4	LC
Swamp cabbage, cooked, boiled, drained	3½ oz.	21	2.2	3.9	LC
Swanson International Entrées Polynesian Style	12¼ oz.	510	19	65	HC
Sweet potatoes:					
Baked in skin	3½ oz.	141	2.1	32.5	HC
Boiled in skin	3½ oz.	114	1.7	26.3	HC
Candied	3½ oz.	168	1.3	34.2	HC
Dehydrated flakes	3½ oz.	95	1.0	22.6	HCR
Tomatoes:					
Canned	3½ oz.	21	1.0	4.3	LC
Cooked, boiled	3½ oz.	26	1.3	5.5	LC
Green, raw	3½ oz.	24	1.2	5.1	LC
Ripe, raw	3½ oz.	22	1.1	4.7	LC
Town House Beets	½ cup	35	1	8.5	LC
Town House Canned Asparagus	½ cup	17.5	2	3	LC
Town House Cream-Style Corn	½ cup	110	2.5	26.5	HCR
Town House Peeled Whole Tomatoes	½ cup	25	1.0	5.5	LC
Town House Pickled Beets	½ cup	70	1	17	LCR
Town House Spinach	½ cup	25	2.5	4	LC
Town House Stewed Tomatoes	½ cup	35	1	9	LC
Town House Sweet Peas	½ cup	70	3.5	12.5	LCR

Food	Quantity	Calories	Protein Grams	Carbo-hydrates Grams	P/C/C Computer
Town House Whole Kernel Yellow Corn	½ cup	80	2	20	HCR
Turnip greens, cooked, boiled, drained	3½ oz.	20	2.2	3.6	LC
Turnip greens, fresh	3½ oz.	28	3.0	5.0	LC
Turnips, cooked, boiled, drained	3½ oz.	23	.8	4.9	LC
Vegetable main dishes, canned, main ingredients (all 3½ oz):					
Peanuts and soya		237	11.7	13.4	HC
Wheat and soy protein		104	16.1	1.2	HP
Wheat and soy protein, soy, or other vegetable oil		150	16.1	5.6	HP
Wheat protein		109	16.3	8.4	HC
Wheat protein, nuts or peanuts		212	20.3	7.1	HP
Wheat protein, vegetable oil		189	19.1	10.4	HP
Vegetables, mixed, frozen	3½ oz.	64	3.2	13.4	LC
Watercress, fresh	3½ oz.	19	2.2	3.0	LC
Yam, fresh	3½ oz.	101	2.1	23.2	HCR

Vegetables and Vegetable Products

Meat, Poultry, Game

Food	Quantity	Calories	Protein Grams	Carbo-hydrates Grams	P/C/C Com-puter
Armour Potted Meat	3½ oz.	229	11.5	–	HC
Armour-Dial Beef Stew	8 oz.	200	10	10	HC
Armour-Dial Corned Beef Hash	7-5/8 oz.	400	17	19	HC
Armour-Dial Potted Meat	3½ oz.	225	11.65	–	HC
Armour-Dial Treet	3½ oz.	348	11.6	3.48	HC
Armour-Dial Vienna Sausage in Beef Stock	3½ oz.	238	8.4	–	HC
Bacon:					
Broiled or fried	3½ oz.	611	30.4	3.2	HC
Canned	3½ oz.	685	8.5	1.0	HC
Canadian, broiled or fried	3½ oz.	277	27.6	.3	HC
B & M Beef Stew	3½ oz.	68	5.7	6.8	LC
B & M Chicken Stew	3½ oz.	57	4.7	6.8	LC
B & M Lamb Stew	3½ oz.	84	7.3	5	LC
Banquet Beef Dinner	11 oz.	312	30.3	20.9	HP
Banquet Beef Pie	8 oz.	409	16.3	40.9	HC
Banquet Chicken Pie	8 oz.	427	15.4	39	HC
Banquet Chopped Beef Dinner	11 oz.	443	18.1	32.8	HC
Banquet Combination Mexican Style	12 oz.	571	22.1	72.1	HCR
Banquet Cooking Bag Beef Chop Suey	7 oz.	121	45	38	HC
Banquet Cooking Bag Chicken a la King	5 oz.	140	59	36	HC
Banquet Cooking Bag Chicken Chow Mein	7 oz.	123	49	43	HCR
Banquet Cooking Bag Chili con Carne w/Beans	8 oz.	310	58	86	HCR
Banquet Cooking Bag Chipped Creamed Beef	5 oz.	126	52	37	HP
Banquet Cooking Bag Meat Loaf	5 oz.	281	55	47	HP
Banquet Cooking Bag Salisbury Steak & Gravy	5 oz.	239	64	29	HP
Banquet Cooking Bag Sliced Beef w/BBQ Sauce	5 oz.	152	55	52	HP
Banquet Cooking Bag Sliced Beef and Gravy	5 oz.	158	79	17	HP
Banquet Cooking Bag Sliced Turkey and Giblet Gravy	5 oz.	129	59	23	HP
Banquet Cooking Bag Sloppy Joe	5 oz.	251	59	144	HP
Banquet Cooking Bag Spaghetti w/Meat Sauce	8 oz.	323	52	135	HCR
Banquet Cooking Bag Tamales w/Sauce	6 oz.	279	34	104	HCR
Banquet Corned Beef Hash Dinner	10 oz.	372	19.9	42.6	HC

Food	Quantity	Calories	Protein Grams	Carbo-hydrates Grams	P/C/C Com-puter
Banquet Dinner Beans and Franks	10¾ oz.	528	16.8	63.1	HCR
Banquet Dinner Beef Chop Suey	12 oz.	282	13.6	38.8	HC
Banquet Dinner Beef Enchilada	12 oz.	479	17	63.6	HC
Banquet Dinner Cheese Enchilada	12 oz.	459	18	58	HC
Banquet Dinner Chicken and Noodles	12 oz.	374	18.7	50.7	HC
Banquet Dinner Chicken Chow Mein	12 oz.	282	12.2	38.8	HC
Banquet Dinner Fried Chicken	10 oz.	530	27.8	48.4	HC
Banquet Dinner Ham	10 oz.	369	16.8	47.7	HC
Banquet Dinner Macaroni and Beef	12 oz.	394	12.6	55.1	HC
Banquet Dinner Meat Loaf	11 oz.	412	20.9	29	HC
Banquet Dinner Mexican Style	16 oz.	608	21.3	73.5	HC
Banquet Dinner Salisbury Steak	11 oz.	390	18.1	24	HC
Banquet Dinner Spaghetti and Meatballs	11½ oz.	450	14.7	62.9	HC
Banquet Dinner Turkey	11 oz.	293	23.4	27.8	HP
Banquet Dinner Veal Parmigiana	11 oz.	421	20.6	42.1	HC
Banquet Entrées Italian Style	11 oz.	446	19.7	44.6	HC
Banquet Entrée Spaghetti w/Meat Sauce	8 oz.	311	13.8	31.3	HC
Banquet Half Fried Chicken	11.6 oz.	960	296	161	HC
Banquet Supper Beef Stew	32 oz.	720	166	329	HC
Banquet Supper Chicken and Dumplings	32 oz.	1306	349	443	HC
Banquet Supper Chicken and Noodles	32 oz.	735	299	245	HC
Banquet Supper Enchilada, Beef w/Cheese and Chili Gravy	32 oz.	259	33	116	HC
Banquet Supper Macaroni and Beef	32 oz.	1092	210	406	HC
Banquet Supper Salisbury Steak	32 oz.	1524	410	186	HC
Banquet Supper Sliced Beef and Gravy	32 oz.	956	422	85	HC
Banquet Supper Sliced Turkey and Giblet Gravy	32 oz.	677	408	88	HP
Banquet Supper Spaghetti and Meatballs	32 oz.	1324	247	416	HC
Banquet Supper Tamales	32 oz.	1488	181	555	HC
Banquet Turkey	8 oz.	415	15.5	40.6	HC
Banquet 2 lb. Supper Beef Chop Suey	32 oz.	554	206	174	HP
Banquet 2 lb. Supper Chicken Chow Mein	32 oz.	563	224	197	HP
Banquet Whole Fried Chicken	26.5 oz.	2195	677	368	HP

Food	Quantity	Calories	Protein Grams	Carbo-hydrates Grams	P/C/C Computer
Beans and frankfurters, canned	3½ oz.	144	7.6	12.6	HC
Beef:					
Bottom round steak cut, broiled or roasted	4 oz.	271	40	0	HP
Chuck cut, blade pot roast	2¾ oz.	218	20.16	0	HP
Chuck cut, blade steak, broiled or roasted	3 oz.	384	21	0	HC
Corned, boneless, medium fat, cooked	2 oz.	223	14	0	HC
Corned, boneless, canned with potato	4 oz.	230	10	9	HP
Dried, chipped, creamed	4 oz.	209	16	6	HP
Eye round roast, broiled or roasted	4½ oz.	340	38	0	HP
Flank steak, stewed or braised	5½ oz.	330	47	0	HC
Fat	1½ oz.	278	3	0	HC
Flank, stew meat, stewed or braised	1½ oz.	68	11	0	LC
Ground chuck, broiled, or roasted	2¾ oz.	363	20	0	HC
Ground round beef, broiled or roasted	5 oz.	450	50	0	HC
Heel of round cut, stewed or braised	4 oz.	261	36	0	HP
Loin end cut, ground sirloin beef, broiled or roasted	4 oz.	346	18	0	HC
Loin end cut, pin bone sirloin steak, broiled or roasted	4½ oz.	260	31	0	HP
Loin end cut, sirloin steak, broiled or roasted	4½ oz.	260	31	0	HP

Food	Quantity	Calories	Protein Grams	Carbo-hydrates Grams	P/C/C Computer
Beef *(cont.)*:					
Plate cut, boiling beef, stewed or braised	3 oz.	107	14	0	HC
Plate cut, short ribs, stewed or braised	½ oz.	32	4	0	LC
Potpie, home prepared, baked	8 oz.	443	16	37	HC
Rib steak, broiled or roasted	3 oz.	246	24	0	HP
Rolled rib roast, broiled or roasted	5 oz.	302	28	0	HC
Rolled flank, stewed or braised	5 oz.	331	47	0	HC
Round steak cut, broiled or roasted	4 oz.	292	32	0	HP
Rump cut, rolled rump, broiled or roasted	3 oz.	197	27	0	HP
Rump roast cut, stewed or braised	4 oz.	188	25	0	HP
Shank cut, knuckle soup bone, broiled, or roasted	¾ oz.	35	6	0	LC
Short loin cut, club steak, broiled or roasted	3 oz.	260	23	0	HP
Short loin cut, filet mignon, broiled or roasted	6 oz.	395	52	0	HC
Short loin cut, Porterhouse steak, broiled or roasted	3½ oz.	242	25	0	HP
Short loin cut, strip or shell steak, broiled or roasted	5½ oz.	380	51	0	HC
Short loin cut, T-bone steak, broiled or roasted	3½ oz.	235	24	0	HP

Food	Quantity	Calories	Protein Grams	Carbo-hydrates Grams	P/C/C Computer
Beef *(cont.)*:					
Short loin cut, Tenderloin steak, broiled or roasted	2½ oz.	148	17	0	HC
Standing rib roast, broiled or roasted	5 oz.	302	28	0	HC
Top round steak cut, broiled or roasted	4 oz.	254	43	0	HP
Vegetable stew, canned	4½ oz.	222	16	20	HCR
Vegetable stew, home recipe with lean beef chuck	4½ oz.	250	17	18	HCR
Brains, all kinds	1½ oz.	60	5	.38	LC
Brains (all kinds, beef, calf, sheep)	3½ oz.	125	10.4	.8	HC
Broadcast Beef Stew	3½ oz.	100	4.6	6.6	LC
Broadcast Corned Beef Hash	3½ oz.	200	8.1	8.9	HC
Celeste Bambino Sausage	4.5 oz.	321	13.7	35.6	HC
Celeste Beef Dinner Ravioli	7.5 oz.	262	12.1	38.9	HC
Celeste Beef Ravioli	4 oz.	259	14.3	37.5	HC
Celeste Deluxe Bambino Pizza	5 oz.	317	13.5	33.4	HC
Celeste Lasagna	8 oz.	413	21.5	23.1	HC
Celeste Sausage Pizza	5.7 oz.	400	17.9	38	HC
Checkerboard Farms Rock Cornish Game Hen	1 hen	500	49	–	HC
Checkerboard Farms Rolled and Tied Turkey Roast	3½ oz.	200	29.78	–	HC
Checkerboard Farms White and Dark Turkey Roast	3½ oz.	213.8	28.42	0.52	HC
Checkerboard Farms White Turkey Roast	3½ oz.	205.04	29.36	0.54	HP
Chicken:					
Breast, edible portion, broiled	2½ oz.	106	18	0	HP
Breast, edible portion, fried	1½ oz.	116	13	1.6	HP
Broiled	7 oz.	281	50	0	HP
Canned, meat only, boned	3½ oz.	198	21.7	0	HP
Capon, flesh and skin, roasted	1 oz.	90	6.8	0	LC
Dark meat without skin, roasted	1 oz.	66	10	0	HP

Food	Quantity	Calories	Protein Grams	Carbo- hydrates Grams	P/C/C Com- puter
Chicken *(cont.)*:					
Drumstick, edible portion, broiled	1½ oz.	64	10	0	HP
Drumstick, edible portion, fried	1½ oz.	62	11	1.5	HP
Hen, stewed	1 oz.	142	11	0	HP
Light meat without skin, roasted	1 oz.	63	11.3	0	HP
Thigh, edible por- tion, broiled	2 oz.	79	13.8	0	HP
Thigh, edible por- tion, fried	1½ oz.	99	12	1.05	HP
Wing, edible portion, broiled	1 oz.	34	6	0	LC
Wing, edible portion, fried	1 oz.	61	6	.62	LCR
Chicken a la king, cooked, home recipe	3½ oz.	191	11.2	5.0	HC
Chicken and noodles, home recipe	3½ oz.	153	9.3	10.7	HC
Chicken fricassee, cooked, home recipe	3½ oz.	161	15.3	3.2	HC
Chicken potpie, frozen	3½ oz.	219	6.7	22.2	HC
Chicken potpie, home-prepared	4 oz.	230	9.6	20.2	HC
Chicken potpie, home-prepared, baked	3½ oz.	235	10.1	18.3	HC
Chili con carne, canned, with beans	3½ oz.	133	7.5	12.2	HC
Chili con carne, canned, with- out beans	3½ oz.	200	10.3	5.8	HC
Chop suey, with meat, canned	3½ oz.	62	4.4	4.2	LC
Chop suey, with meat, home recipe	3½ oz.	120	10.4	5.1	HC
Chow mein, chicken, without noodles, canned	3½ oz.	38	2.6	7.1	LC
Chow mein, chicken, without noodles, home recipe	3½ oz.	102	12.4	4.0	LCR
Desiccated liver	1 tbsp.	40	8.5	0	HP
Duck:					
Domestic	3½ oz.	165	21.4	0	HC
Roasted	12 oz.	109	8	0	HC
Wild	3½ oz.	138	21.3	0	HC
Eckrich Beef Bologna Pickle Loaf	2 slices	130	5	2	HC
Eckrich Beef Fun Frankfurters	2 franks	310	11	4	HC
Eckrich Beef Smokettes Sausage	2 sausages	170	6	2	HC

Food	Quantity	Calories	Protein Grams	Carbo-hydrates Grams	P/C/C Computer
Eckrich Fun Franks	2 franks	240	8	3	HC
Eckrich Garlic Bologna	2 slices	190	6	3	HC
Eckrich Garlic Ring Bologna	2 oz.	190	6	3	HC
Eckrich Gourmet Loaf	2 slices	80	8	4	LCR
Eckrich Honey Style Loaf	2 slices	90	8	4	LCR
Eckrich Jumbo Fun Frankfurters	1 frank	190	6	3	HC
Eckrich Lunch Bologna	2 oz.	190	6	3	HC
Eckrich Old-Fashioned Loaf	2 slices	150	7	4	HC
Eckrich Pickle Loaf	2 slices	170	6	3	HC
Eckrich Pickled Ring Bologna	2 oz.	190	6	3	HC
Eckrich Polska Kielbasa	2 oz.	200	6	1	HC
Eckrich Polska Kielbasa Links	2 links	310	12	3	HC
Eckrich Polska Kielbasa Sausage	2 oz.	190	7	2	HC
Eckrich Ring Bologna	2 oz.	190	6	3	HC
Eckrich Sandwich Bologna	2 slices	180	7	3	HC
Eckrich Skinless Smoked Sausage	2 oz.	190	7	2	HC
Eckrich Sliced Bologna	2 slices	190	6	3	HC
Eckrich Smoked Sausage	2 oz.	210	7	1	HC
Eckrich Smokees	2	230	8	2	HC
Eckrich Smok-Y-Links	2 links	170	6	2	HC
Eckrich Smok-Y-Links, Maple Flavor	2 links	170	6	2	HC
Eckrich Thick-Sliced Bologna	2 slices	170	6	3	HC
Franco-American Macaroni 'n Beef in Tomato Sauce	7½ oz.	220	8	27	HCR
Franco-American Spaghetti and Beef in Tomato Sauce	7½ oz.	240	11	34	HCR
Franco-American Spaghetti w/Meatballs	7¼ oz.	230	9	22	HCR
Franco-American Spaghetti O's w/Little Meatballs	7½ oz.	230	10	24	HCR
Franco-American Spaghetti O's w/Sliced Franks	7½ oz.	260	9	28	HCR
Gizzard, chicken, simmered	3½ oz.	148	27.0	.7	HP
Gizzard, goose	3½ oz.	139	21.4	0	HP
Gizzard, turkey, simmered	3½ oz.	196	26.8	1.1	HP
Goose, roasted	3½ oz.	322	28	0	HC
Guinea hen	3½ oz.	158	23.4	0	HP
Ham croquette	3½ oz.	251	16.3	11.7	HC
Hamburger Helper Beef Noodle	1/5 pkg.	150	5	26	HCR
Hamburger Helper Cheeseburger	1/5 pkg.	180	6	26	HCR
Hamburger Helper Hamburger Stew	1/5 pkg.	110	3	23	HCR
Hamburger Helper Hash	1/5 pkg.	130	3	25	HCR
Hamburger Helper Lasagna	1/5 pkg.	160	6	30	HCR
Hamburger Helper Potato Stroganoff	1/5 pkg.	160	3	30	HCR
Hamburger Helper Rice Oriental	1/5 pkg.	140	3	30	HCR

Food	Quantity	Calories	Protein Grams	Carbo-hydrates Grams	P/C/C Computer
Heart, beef, lean, braised	3½ oz.	188	31	.7	HP
Heinz Beans and Franks in Tomato Sauce	3½ oz.	158	7.1	14.7	HC
Heinz Beef Goulash	3½ oz.	113	6.3	9.9	HC
Heinz Beef Stew	3½ oz.	95	6.2	7.9	LC
Heinz Chicken Noodle w/Gravy	3½ oz.	75	4.4	7.3	LC
Heinz Chicken Stew w/Dumplings	3½ oz.	95	4.4	10.4	LCR
Heinz Noodles w/Beef and Sauce	3½ oz.	85	4.1	9.9	LCR
Heinz Spaghetti and Franks	3½ oz.	128	4.5	11.7	HC
Heinz Spaghetti w/Meat Sauce	3½ oz.	86	4.3	10.9	LC
Hellman's Sandwich Spread	1 tbsp.	60	–	2	LC
Hunt's Barbecue-Flavored Manwich	3½ oz.	65.8	1.48	15.8	LC
Hunt's Big John Beans and Fixin's	1 pkg.	769.5	29.2	128.8	HC
Hunt's Hawaiian Skillet	3½ oz.	163.3	2.78	25.8	HC
Hunt's Lasagne Skillet	1 pkg.	820	41.3	112.8	HCR
Hunt's Mexicana Skillet	3½ oz.	152	4.24	27.4	HC
Hunt's Oriental Skillet	3½ oz.	132.9	4.75	22.9	HC
Hunt's Pizzeria Skillet	1 pkg.	507	24.82	77.5	HCR
Hunt's Regular Manwich	3½ oz.	55.3	1.6	13.2	LC
Hunt's Skillet Stroganoff	3½ oz.	153.6	3.66	18.1	HC
Hygrade Ball Park Frankfurters	1 frank	119	6.75	–	HC
Jeno's Pepperoni Pizza	7.5 oz.	630	22	82	HC
Jeno's Regular Pizza	7 oz.	480	14	80	HC
Jeno's Sausage Pizza	7.5 oz.	560	72	70	HC
Kellogg's Chicken Spread	2 oz.	141	6.5	6.3	HC
Kellogg's Ham Spread	2 oz.	124	5.7	5.8	HC
Kidneys:					
Beef, braised	5 oz.	390	51	1.24	HC
Beef, cooked, braised	3½ oz.	252	33.0	.8	HP
Calf	3½ oz.	113	16.6	.1	HP
Lamb	3½ oz.	105	16.8	.9	HP
La Choy Beef Chow Mein	1/3 cup	47.17	4.34	6.35	LCR
La Choy Beef Chow Mein	1 cup	77	9.3	4.4	LCR
La Choy Beef Chow Mein Bi-Pack	1 cup	90	9.3	7.4	LCR
La Choy Chicken Chow Mein	1/3 cup	84.14	3.75	12.35	LCR
La Choy Chicken Egg Rolls	3½ oz.	243.13	7.56	35.61	HCR
La Choy Fried Rice w/Chicken	1 cup	137	5.0	25.4	HCR
La Choy Fried Rice w/Meat	2/3 cup	130.14	5.87	26.27	HCR
La Choy Skillet Dinners Chicken Chow Mein	3½ oz.	42	7.75	3.78	LC
La Choy Skillet Dinners Pepper Steak	3½ oz.	91	8.80	3.41	LC
La Choy Skillet Dinners Sukiyaki	3½ oz.	110	9.55	3.90	HC

Food	Quantity	Calories	Protein Grams	Carbo-hydrates Grams	P/C/C Computer
La Choy Skillet Dinners Sweet and Sour	3½ oz.	139	13.02	15.62	HC
La Choy Skillet Dinners Teriyaki	3½ oz.	133	10.77	3.19	HC
La Choy Sweet and Sour Pork	2/3 cup	99.08	4.43	21.98	HCR
Lamb:					
Breast	2 oz.	147	9.24	0	HC
Leg of lamb	3½ oz.	192	27.6	0	HP
Rib chop	1½ oz.	119	10.5	0	HP
Shoulder cut	2 oz.	102	13	0	HP
Shoulder cut, cubes for shish-kebab	2 oz.	51	6.7	0	LC
Stew	1 oz.	65	4.12	0	LC
Liver:					
Beef, fried	1½ oz.	86	8.8	5.6	LC
Calf, fried	1½ oz.	74	8.1	1.7	LC
Chicken, simmered	3½ oz.	148	23.85	2.79	HP
Lamb, broiled	1 oz.	86	9.3	4.3	LC
Pork, fried	1 oz.	85	8.8	3.8	LC
Morton Beans and Franks and Sauce	12 oz.	554.40	19.01	81.28	HC
Morton Beef Dish	11 oz.	292.11	28.72	21.99	HC
Morton Beef Pie	8 oz.	369.29	14.87	35.10	HC
Morton Casserole Spaghetti and Meat	8 oz.	288	13.86	28.91	HC
Morton Chicken Dish	11 oz.	481.61	26.19	37.24	HC
Morton Chicken and Dumplings	12 oz.	371.75	22.96	30.70	HC
Morton Chicken Pie	8 oz.	445	17.33	35.35	HC
Morton Ham Dinner	11 oz.	429.28	18.45	49.08	HC
Morton Macaroni and Beef	11 oz.	365.28	11.46	56.33	HC
Morton Meat Loaf	11 oz.	397.56	24.3	27.69	HC
Morton Salisbury Steak	11 oz.	383.90	21.04	24.73	HC
Morton Spaghetti and Meatballs	11 oz.	367	13.36	55.91	HCR
Morton 3-Course Chicken and Dumplings	16.5 oz.	768.10	30.81	90.71	HC
Morton 3-Course Fried Chicken	17 oz.	867.61	33.46	73.18	HC
Morton 3-Course Meat Loaf	17 oz.	651.89	32.71	69.37	HC
Morton 3-Course Salisbury Steak	17 oz.	611.74	26.86	66.64	HC
Morton 3-Course Sliced Beef	17 oz.	562.72	34.75	60.08	HC
Morton 3-Course Sliced Turkey	17 oz.	634.23	25.72	19.41	HC
Morton Turkey	8 oz.	410.76	17.02	34.77	HC
Morton Turkey	12 oz.	444.91	26.37	38.60	HC
Mrs. Paul's Beef and Cabbage	1 oz.	36	4.40	8.08	LC
Mrs. Paul's Beef and Eggplant	1 oz.	35	4.28	9.33	LC
Mrs. Paul's Beef and Green Peppers	1 oz.	32.2	4.47	9.13	LC

Food	Quantity	Calories	Protein Grams	Carbo-hydrates Grams	P/C/C Com-puter
Mrs. Paul's Beef and Zucchini	1 oz.	31.8	5.22	8.86	LC
Nu Made Sandwich Spread	1 tbsp.	65	0	4	LC
Oregon Freeze Dry Beans and Franks	3½ oz.	450	22	48.5	HC
Oregon Freeze Dry Beef Chop Suey	3½ oz.	420	25.5	52.5	HC
Oregon Freeze Dry Beef, Diced, Cooked	3½ oz.	470	78	0.5	HC
Oregon Freeze Dry Beef and Potatoes	3½ oz.	480	22	44	HC
Oregon Freeze Dry Beef w/Rice	3½ oz.	410	21.5	62	HC
Oregon Freeze Dry Beef Stew	3½ oz.	440	33.5	43	HC
Oregon Freeze Dry Chicken Chop Suey	3½ oz.	430	20	55	HC
Oregon Freeze Dry Chicken Pilaf	3½ oz.	460	16.5	60	HC
Oregon Freeze Dry Chicken Salad	3½ oz.	460	34.5	33	HC
Oregon Freeze Dry Chicken Stew	3½ oz.	430	31	42.5	HC
Oregon Freeze Dry Rice and Chicken	3½ oz.	450	13	66	HCR
Oregon Freeze Dry Sausage Patties	3½ oz.	540	49.5	5.5	HC
Oregon Freeze Dry Vegetable Stew w/Beef	3½ oz.	440	24.5	52	HC
Oscar Mayer Bacon	3½ oz.	605	22	1.3	HC
Oscar Mayer Beef Frankfurters	3½ oz.	315	11	–	HC
Oscar Mayer Bologna	3½ oz.	315	11	3	HC
Oscar Mayer Braunschweiger	3½ oz.	380	13	0.6	HC
Oscar Mayer Cotto Salami	3½ oz.	235	15	1.6	HC
Oscar Mayer Ham Jubilee	3½ oz.	130	21	–	HC
Oscar Mayer Meat Frankfurters	3½ oz.	315	11	2.8	HC
Oscar Mayer Smokie Links	3½ oz.	305	13	2.3	HC
Pancreas, medium-fat, beef	3½ oz.	283	13.5	0	HC
Pantry Pride Beef Frankfurters	3½ oz.	309	13	3	HC
Pantry Pride Bologna	3½ oz.	304	11	–	HC
Pantry Pride Liverwurst	3½ oz.	319	26	–	HC
Pantry Pride Meat Frankfurters	3½ oz.	309	13	–	HC
Pâté de foie gras, canned	3½ oz.	462	11.4	4.8	HC
Pâté de foie gras, canned	1 tbsp.	92.4	2.28	.96	LCR
Pheasant, total edible	3½ oz.	151	24.3	0	HP
Pigs' feet, pickled	3½ oz.	199	16.7	0	HC
Pork:					
And beef (chopped together)	3½ oz.	336	15.6	0	HC
And gravy, canned (90% pork, 10% gravy)	3½ oz.	256	16.4	6.3	HC

Food	Quantity	Calories	Protein Grams	Carbo-hydrates Grams	P/C/C Com-puter
Butt, blade steak	2 oz.	150	15.7	0	HC
Cured ham butt	2 oz.	123	15	.2	HC
Cured ham, canned	3½ oz.	193	18.3	.9	HP
Cured ham shank	1½ oz.	91	10	.1	LC
Ham butt slice	3½ oz.	374	23	0	HP
Ham center slice	3½ oz.	374	23	0	HP
Ham roast	1½ oz.	126	20	0	HP
Loin, center cut	3½ oz.	362	24.5	0	HC
Loin, crown roast	1½ oz.	101	11.76	0	HP
Loin, pork chop	1½ oz.	111	13	0	HP
Loin, rib chop	1½ oz.	101	11.76	0	HP
Loin, roast ham end	3½ oz.	380	25.72	0	HC
Loin, roast shoulder	3½ oz.	362	24.5	0	HC
Loin, sirloin roast	1 oz.	54	7.1	0	LC
Loin, tenderloin	1½ oz.	114	13	0	HP
Picnic, rolled shoulder	1½ oz.	116	12.2	0	HP
Picnic, shoulder hock	1½ oz.	116	12.2	0	HP
Rolled ham roast	1½ oz.	126	20	0	HP
Sausage, canned	3½ oz.	415	13.8	38.4	HC
Sausage, links or bulk	3½ oz.	476	18.1	Trace	HC
Side, salt pork, fried	1½ oz.	341	6	0	HC
Spareribs	1½ oz.	105	5	0	LC
Quail, total edible	3½ oz.	168	25.0	0	HP
R & R Canned Chicken a la King	3½ oz.	121	8.9	6.4	LP
R & R Chicken Fricassee	3½ oz.	114	8.9	6.7	LP
Rabbit, stewed	1½ oz.	82.08	11.13	0	LC
Salami	3½ oz.	311	17.5	1.4	HC
Salt pork	3½ oz.	783	3.9	0	HC
Sausage, cold cuts, and luncheon meats:					
Blood sausage or blood pudding	3½ oz.	394	14.1	.3	HC
Bockwurst	3½ oz.	264	11.3	.6	HC
Bologna	3½ oz.	304	12.1	1.1	HC
Braunschweiger	3½ oz.	319	14.8	2.3	HC
Brown-and-serve sausage	3½ oz.	422	16.5	2.8	HC
Capicola or Capacola	3½ oz.	499	20.2	0	HC
Cervelat, dry	3½ oz.	451	24.6	1.7	HC
Cervelat, soft	3½ oz.	307	18.6	1.6	HC
Country-style sausage	3½ oz.	345	15.1	0	HC
Deviled ham, canned	3½ oz.	351	13.9	0	HC
Frankfurters, canned	3½ oz.	221	13.4	.2	HC

Food	Quantity	Calories	Protein Grams	Carbo-hydrates Grams	P/C/C Com-puter
Sausage, cold cuts, and luncheon meats *(cont.)*:					
Frankfurters, fresh	3½ oz.	309	12.5	1.8	HC
Headcheese	3½ oz.	268	15.5	1.0	HC
Knockwurst	3½ oz.	278	14.1	2.2	HC
Liverwurst, fresh	3½ oz.	307	16.2	1.8	HC
Liverwurst, smoked	3½ oz.	319	14.8	2.3	HC
Luncheon meat:					
Boiled ham	3½ oz.	234	19.0	0	HC
Meat loaf	3½ oz.	200	15.9	3.3	HC
Meat, potted (includes beef, chicken, turkey)	3½ oz.	248	17.5	0	HC
Minced ham	3½ oz.	228	13.7	4.4	HC
Mortadella	3½ oz.	315	20.4	.6	HC
Polish-style sausage	3½ oz.	304	15.7	1.2	HC
Pork, cured ham or shoulder	3½ oz.	294	15.0	1.3	HC
Scrapple	3½ oz.	215	8.8	14.6	HC
Souse	3½ oz.	181	13.0	1.2	HC
Squab (pigeon), total edible	3½ oz.	279	18.6	0	HC
Swanson Beef Dinner	11½ oz.	370	31	34	HC
Swanson Beef Pie	8 oz.	430	12	43	HC
Swanson Beef Stew	7½ oz.	190	13	18	HC
Swanson Boned Turkey w/Broth	2½ oz.	110	17	—	HP
Swanson Canned Boned Chicken w/Broth	2½ oz.	110	15	1	HC
Swanson Canned Chicken a la King	3½ oz.	190	12	9	HC
Swanson Chicken and Dumplings	7½ oz.	230	12	18	HC
Swanson Chicken Pie	8 oz.	460	14	44	HC
Swanson Chicken Spread	2 oz.	140	8	4	HC
Swanson Chicken Stew	7½ oz.	180	10	18	HC
Swanson Chili con Carne	7¾ oz.	300	14	28	HC
Swanson Deep Dish Beef Pie	16 oz.	770	30	65	HC
Swanson Deep Dish Chicken Pie	16 oz.	750	31	62	HC
Swanson Deep Dish Turkey	16 oz.	760	29	60	HC
Swanson Dinner Beans and Franks	11¼ oz.	550	16	75	HC
Swanson Dinner Chicken and Noodles	10¼ oz.	380	14	46	HC
Swanson Dinner Fried Chicken	11½ oz.	570	28	48	HC

Food	Quantity	Calories	Protein Grams	Carbo-hydrates Grams	P/C/C Computer
Swanson Dinner Italian Style	13½ oz.	510	18	58	HC
Swanson Dinner Macaroni and Beef	12 oz.	410	12	56	HC
Swanson Dinner Salisbury Steak	11½ oz.	500	20	40	HC
Swanson Dinner Spaghetti and Meatballs	12½ oz.	320	14	44	HC
Swanson Dinner Swiss Steak	10 oz.	380	22	36	HC
Swanson Dinner Veal Parmigiana	12¼ oz.	520	23	47	HC
Swanson Ham Dinner	10¼ oz.	380	19	47	HC
Swanson Hungry Man Boneless Chicken	19 oz.	770	43	69	HC
Swanson Hungry Man Fried Chicken	15¾ oz.	960	45	78	HC
Swanson Hungry Man Salisbury Steak	17 oz.	930	42	60	HC
Swanson Hungry Man Turkey	19 oz.	690	42	78	HC
Swanson International Dinner Beef Enchilada	15 oz.	570	19	72	HC
Swanson International Dinner Mexican Style	16 oz.	700	22	75	HC
Swanson International Entrées Chicken Livers	4 oz.	120	20	4	HP
Swanson International Entrées Chicken w/Potatoes	7 oz.	530	23	42	HC
Swanson International Entrées Chinese Style	8½ oz.	200	10	25	HC
Swanson International Entrées English Style	5 oz.	300	14	29	HC
Swanson International Entrées Fried Chicken	4 oz.	300	21	13	HC
Swanson International Entrées Italian Style	9½ oz.	280	10	44	HC
Swanson International Entrées Mexican Style	10 oz.	450	16	43	HC
Swanson Supper Noodles and Beef	32 oz.	735	200	408	HC
Swanson 3-Course Beef Dish	15 oz.	540	30	58	HC
Swanson 3-Course Fried Chicken	15 oz.	670	31	64	HC
Swanson 3-Course Meat Loaf	16½ oz.	540	27	53	HC
Swanson 3-Course Mexican Style	18 oz.	620	22	74	HC
Swanson 3-Course Salisbury Steak	16 oz.	510	25	48	HC
Swanson 3-Course Turkey	16 oz.	550	30	60	HC
Swanson Turkey	8 oz.	450	13	40	HC
Swanson TV Brand Entrées Fried Chicken w/Potatoes	7oz.	380	20	25	HC
Swanson TV Brand Entrées Meat Loaf w/Tomato Sauce and Potatoes	9 oz.	320	19	27	HC

Meat, Poultry, Game

Food	Quantity	Calories	Protein Grams	Carbo-hydrates Grams	P/C/C Com-puter
Swanson TV Brand Entrées Meat-balls w/Gravy and Potatoes	9¼ oz.	330	18	26	HC
Swanson TV Brand Entrées Salisbury Steak w/Potatoes	5½ oz.	370	15	28	HC
Swanson TV Brand Entrées Spaghetti in Tomato Sauce w/Veal	8¾ oz.	260	20	27	HC
Swanson TV Brand Entrées Turkey/Gravy/Dressing/ Potatoes	8¾ oz.	260	20	27	HC
Sweetbreads (thymus):					
Beef	3½ oz.	320	25.9	0	HC
Calf	3½ oz.	168	32.6	0	HC
Lamb	3½ oz.	175	28.1	0	HC
Swift Boneless Breast Turkey Roast	3 slices	143	42	0.7	HP
Swift Boneless White and Dark Turkey Roast	3 slices	152	41	0.7	HC
Swift Brown 'n Serve Sausage, Bacon 'n Sausage	3 links	216	15	1.2	HC
Swift Brown 'n Serve Kountry Kitchen Sausage	3 links	258	12	0.9	HC
Swift Brown 'n Serve Premium Sausage	3½ oz.	388	14.6	2.6	HC
Swift Butterball Turkey	3½ oz.	219	28.4	–	HC
Swift Butterball Turkey, White Meat	2 slices	173	47	–	HC
Swift Hostess Ham, Premium Cooked	1 slice	141	30	1.6	HC
Swift Premium Ham	2 slices	221	28	0.8	HC
Thuringer	3½ oz.	307	18.6	1.6	HC
Tongue, beef, braised	1 oz.	51	4.51	.08	LC
Tripe, beef, commercial	3½ oz.	100	19.1	0	HP
Tripe, beef, pickled	3½ oz.	62	11.8	0	LC
Turkey	1½ oz.	75	13	0	LC
Turkey, canned, meat only	3½ oz.	202	20.9	0	HC
Turkey potpie, home-baked	3½ oz.	237	10.4	18.5	HC
Turkey potpie, frozen	3½ oz.	197	5.8	20.1	HC
Underwood Chicken Spread	2 oz.	127.4	8.96	2.1	HC
Underwood Corned Beef Spread	2 oz.	110.8	7.64	0.57	HC
Underwood Deviled Ham Spread	2 oz.	195.4	7.6	0	HC
Underwood Liverwurst Spread	2 oz.	185	8.6	2.2	HC
Van Camp Beanee-Weanee Meat Dish	1 cup	315	16.6	27.6	HC
Van Camp Beef Stew	1 cup	204	15	18.4	HC
Van Camp Corned Beef Hash	1 cup	416	20.6	24.6	HC

Food	Quantity	Calories	Protein Grams	Carbo-hydrates Grams	P/C/C Computer
Van Camp Potted Meat	1 cup	544	38.4	–	HC
Van Camp Spaghetti w/Meatballs	1 cup	226	10.6	25	HC
Van Camp Vienna Sausage	4 oz.	288	17	0.4	HC
Veal:					
Breast, moist-cooked	1 oz.	82	10	0	HP
Breast, stew meat	4 oz.	121	8.8	3.6	HC
Frenched rib chop	1½ oz.	107	10	0	HP
Loin chop	1 oz.	70.2	8	0	HP
Loin or kidney chop	1 oz.	82	10	0	HP
Rib crown roast	1½ oz.	107	10	0	LC
Rib roast	1½ oz.	102	10	0	LC
Rolled shoulder	1½ oz.	163	10	0	HC
Round, rolled rump roast	1½ oz.	84	14	0	LC
Round, scallops	1½ oz.	95	12	0	HP
Round roast	1½ oz.	121	15	0	HP
Round rump roast	1½ oz.	84	14	0	LC
Round steak (veal cutlet)	2 oz.	151	19	0	HP
Shank, patties	2 oz.	113	15	0	HP
Shoulder, blade roast	1½ oz.	163	10	0	HC
Sirloin steak	3½ oz.	172	27	0	HP
Venison, lean meat only, fresh	3½ oz.	126	21	0	HP
Vienna sausage, canned	3½ oz.	240	14.0	.3	HC
Weight Watchers Beef Sirloin Dinner	16 oz.	595	0.8	0.8	HC
Weight Watchers Beef Steak w/Carrots Luncheon	10 oz.	412	8.6	2.8	HC
Weight Watchers Beef Steak w/Cauliflower Luncheon	11 oz.	431	8.4	3.6	HC
Weight Watchers Chicken Creole Luncheon	12 oz.	225	0.9	0.3	HC
Weight Watchers Chicken Livers Luncheon	11½ oz.	234	0.95	3.9	HC
Weight Watchers Chopped Chicken Liver Luncheon	5 oz.	229	2.6	9.3	HC
Weight Watchers Turkey Dinner	18 oz.	302	7.9	2.4	HC
Weight Watchers Veal Stuffed Pepper Dinner	12 oz.	335	0.8	0.2	HC
Weight Watchers White Meat Chicken Luncheon	10 oz.	284	12.6	6.3	HC

Fish and Seafood

Food	Quantity	Calories	Protein Grams	Carbo-hydrates Grams	P/C/C Computer
Abalone, raw	3½ oz.	98	18.7	3.4	HP
Albacore, raw	3½ oz.	177	25.3	0	HP
Alewife, canned	3½ oz.	141	16.2	0	HP
Anchovy, pickled	3½ oz.	176	19.2	.3	HP
Barracuda, Pacific, raw	3½ oz.	113	21.0	0	HP
Bass, black sea	3½ oz.	259	16.2	11.4	HP
Bass, striped	3½ oz.	196	21.5	6.7	HP
Banquet Dinner Haddock	8-3/4 oz.	419	21.3	45.4	HC
Banquet Dinner Ocean Perch	8-3/4 oz.	434	19.1	49.8	HC
Banquet Tuna	8 oz.	434	14.1	42.7	HC
Bluefish, baked or broiled	3½ oz.	159	26.2	0	HP
Bluefish, fried	3½ oz.	205	22.7	4.7	HP
Bonito fish, baked or broiled, fried	3½ oz.	168	24.0	0	HP
Buffalo fish	3½ oz.	113	17.5	0	HP
Bullhead, black	3½ oz.	84	16.3	0	HP
Bumble Bee Baby Shrimp	3½ oz.	70	14.23	0.7	HP
Bumble Bee Clams	3½ oz.	51.33	7.78	2.8	LC
Bumble Bee Shelled Oysters	3½ oz.	75.25	8.4	4.81	LC
Bumble Bee Sockeye Red Salmon	3½ oz.	125.13	14.7	—	HP
Bumble Bee Tuna	3½ oz.	146.13	21.44	—	HP
Burbot fish	3½ oz.	92	37.0	0	HP
Butterfish	3½ oz.	169	18.1	0	HP
Carp	3½ oz.	115	18.0	0	HP
Catfish, freshwater	3½ oz.	103	17.6	0	HP
Caviar, sturgeon, pressed	3½ oz.	316	34.4	4.9	HP
Chicken of the Sea Light Chunk Tuna in Oil	3½ oz.	220	25	1	HP
Chicken of the Sea Solid Light Tuna in Oil	3½ oz.	260	26	1	HP
Chicken of the Sea White Solid Tuna in Oil	3½ oz.	290	24	1	HP
Chicken of the Sea White Solid Tuna in Water	3½ oz.	100	27	1	HP
Chub	3½ oz.	145	15.3	0	HP
Clam fritters	3½ oz.	311	11.4	30.9	HC
Clams, canned, meat and liquid	3½ oz.	52	7.9	2.8	LC
Clams, raw, meat and liquid	3½ oz.	54	8.6	2.0	LC
Codfish:					
Broiled	3½ oz.	170	28.5	0	HP
Canned	3½ oz.	85	19.2	0	LC
Dehydrated, lightly salted	3½ oz.	375	81.8	0	HP
Dried, salted	3½ oz.	130	29.0	0	HP
Crab:					
Canned	3½ oz.	101	17.4	1.1	HP

Food	Quantity	Calories	Protein Grams	Carbo-hydrates Grams	P/C/C Computer
Crab *(cont.)*:					
Cooked, steamed	3½ oz	93	17.3	.5	HP
Deviled	3½ oz.	188	11.4	13.3	HC
Imperial	3½ oz.	147	14.6	3.9	HC
Crappie, white	3½ oz.	79	16.8	0	HP
Crayfish	3½ oz.	72	14.6	1.2	HP
Croaker:					
Atlantic	3½ oz.	133	24.3	0	HP
White	3½ oz.	84	18.0	0	HP
Yellowfin	3½ oz.	89	19.2	0	HP
Del Monte Pink Salmon	3½ oz.	121.19	19.56	–	HP
Del Monte Sardines in Tomato Sauce	3½ oz.	138	17.8	1.4	HP
Del Monte Sockeye Red Salmon	3½ oz.	136.94	16.19	–	HP
Del Monte Tuna in Oil	3½ oz.	163.63	17.07	–	HP
Del Monte Tuna Albacore in Oil	3½ oz.	133.88	19.16	–	HP
Dogfish, spiny	3½ oz.	156	17.6	0	HP
Dolly varden	3½ oz.	144	19.9	0	HP
Drum, freshwater	3½ oz.	121	17.3	0	HP
Drum, redfish	3½ oz.	80	18.0	0	HP
Eel, American	3½ oz.	233	15.9	0	HC
Eel, smoked	3½ oz.	330	18.6	0	HC
Eulachon (smelt)	3½ oz.	118	14.6	0	HP
Finan haddie (smoked haddock)	3½ oz.	103	23.2	0	HP
Fish cakes, fried	3½ oz.	172	14.7	9.3	HC
Fish cakes, frozen, fried, reheated	3½ oz.	270	9.2	17.2	HC
Fish flakes, canned	3½ oz.	111	24.7	0	HP
Fish loaf	3½ oz.	124	14.1	7.3	HP
Fish sticks, frozen, cooked	3½ oz.	176	16.6	6.5	HC
Flatfishes (flounders, soles)	3½ oz.	79	16.7	0	HP
Flounder, baked	3½ oz.	202	30.0	0	HP
Frog legs	3½ oz.	73	16.4	0	HP
Grouper	3½ oz.	87	19.3	0	HP
Haddock, cooked, fried	3½ oz.	165	19.6	5.8	HP
Haddock, smoked, canned	3½ oz.	103	23.2	0	HP
Hake (whiting)	3½ oz.	74	16.5	0	HP
Halibut:					
Atlantic and Pacific, cooked, broiled	3½ oz.	171	25.2	0	HC
California	3½ oz.	97	19.8	0	HP
Greenland	3½ oz.	146	16.4	0	HP
Smoked	3½ oz.	224	20.8	0	HC
Herring:					
Canned, in tomato sauce	3½ oz.	176	15.8	3.7	HP

Food	Quantity	Calories	Protein Grams	Carbo-hydrates Grams	P/C/C Computer
Herring *(cont.)*:					
Fresh	3½ oz.	176	17.3	0	HP
Pickled, Bismarck type	3½ oz.	223	20.4	0	HP
Salted or brined	3½ oz.	218	19.0	0	HP
Smoked, kippered	3½ oz.	211	22.2	0	HP
Inconnu (sheefish)	3½ oz.	146	19.9	0	HP
Jack mackerel	3½ oz.	143	21.6	0	HP
Kingfish (whiting)	3½ oz.	105	18.3	0	LC
La Choy Lobster Egg Rolls	3½ oz.	203	7.90	32.88	HCR
La Choy Meat and Shrimp Egg Rolls	3½ oz.	198.37	6.14	32.49	HCR
La Choy Shrimp Chow Mein	3½ oz.	32.91	2.61	6.10	LC
La Choy Shrimp Chow Mein	1 cup	75	8.2	4.7	LC
La Choy Shrimp Chow Mein Bi-Pack	1 cup	110	6.3	8.2	LP
La Choy Shrimp Egg Rolls	3½ oz.	203	7.90	32.88	HCR
La Choy Skillet Dinner Cantonese Seafood	3½ oz.	32	6.04	3.59	LP
Lake herring (cisco)	3½ oz.	96	17.7	0	HP
Lake trout	3½ oz.	168	18.3	0	HP
Lingcod	3½ oz.	84	17.9	0	HP
Lobster:					
Newburg	3½ oz.	194	18.5	5.1	HP
Northern, canned or cooked	3½ oz.	95	18.7	.3	HP
Salad	3½ oz.	110	10.1	2.3	LCR
Louisiana Canned Shrimp	3½ oz.	117	24.2	0.7	HP
Louisiana Shelled Oysters	3½ oz.	66	8.5	4.9	LC
Mackerel:					
Atlantic, broiled with butter or margarine	3½ oz.	236	21.8	0	HP
Atlantic, canned	3½ oz.	183	19.3	0	HP
Pacific, broiled with butter or margarine	3½ oz.	220	21.9	0	HP
Pacific, canned	3½ oz.	180	21.1	0	HP
Salted	3½ oz.	305	18.5	0	HP
Smoked	3½ oz.	219	23.8	0	HP
Menhaden, Atlantic, canned	3½ oz.	172	18.7	0	HP
Morton Dinner Fish	8.75 oz.	167.7	42.36	375.36	HCR
Morton Dinner Shrimp	7.75 oz.	378.93	20.96	37.34	HC
Morton Tuna	8 oz.	385.37	14.81	35.78	HC
Mrs. Paul's Buttered Fish Fillet	1 oz.	37.6	17.01	1.10	LC

Food	Quantity	Calories	Protein Grams	Carbo-hydrates Grams	P/C/C Computer
Mrs. Paul's Clam Sticks	1 oz.	57.2	9.20	26.07	HCR
Mrs. Paul's Combination Seafood Platter	1 oz.	57.4	9.20	22.58	HCR
Mrs. Paul's Deviled Crabs	1 oz.	57.5	9.3	20.17	HCR
Mrs. Paul's Fish and Chips	1 oz.	50.6	8.85	23.56	HCR
Mrs. Paul's Fried Clams	1 oz.	101.1	13.50	28.80	HCR
Mrs. Paul's Fried Fish Fillets	1 oz.	42	13.62	15.54	HP
Mrs. Paul's Fried Scallops	1 oz.	58.8	13.06	22.94	HCR
Mrs. Paul's Fried Shrimp	1 oz.	64.6	11.91	19.15	HP
Mrs. Paul's Frozen Deviled Crabs	1 cake	162	7	17	HC
Mrs. Paul's Frozen Fish Cakes	2 cakes	212	10	23	HC
Mrs. Paul's Frozen Fish Fillets	2 fillets	212	12	20	HC
Mrs. Paul's Frozen Fish Sticks	4 sticks	156	9	16	HC
Mrs. Paul's Frozen Fried Scallops	3½ oz.	210	12	24	HC
Mrs. Paul's Frozen Shrimp Cakes	1 cake	155	8	17	HC
Mrs. Paul's Shrimp Cakes	1 oz.	52.5	9.81	21.34	HCR
Mrs. Paul's Thin Clam Cakes	1 oz.	58.6	10.55	22.48	HCR
Mrs. Paul's Thin Crab Cakes	1 oz.	63.2	9.25	21.96	HCR
Mrs. Paul's Thin Fish Cakes	1 oz.	55.2	9.75	25.21	HCR
Mrs. Paul's Thin Shrimp Cakes	1 oz.	63	10.51	21.61	HP
Mullet, striped	3½ oz.	146	19.6	0	HC
Mussels, meat and liquid	3½ oz.	66	9.6	3.1	HP
Mussels, meat only	3½ oz.	95	14.4	3.3	HP
Ocean perch, (Atlantic redfish), cooked, fried	3½ oz.	227	19.0	6.8	HC
Ocean perch, frozen, breaded, fried, reheated	3½ oz.	319	18.9	16.5	HC
Old Vienna Gefilte Fish	3½ oz.	65.5	8.01	4.82	HCR
Oregon Freeze Dry Shrimp Creole	3½ oz.	430	18.5	57.5	HC
Oregon Freeze Dry Tuna Salad	3½ oz.	520	35	27.5	HC
Oysters:					
Canned, solids and liquid	3½ oz.	76	8.5	4.9	LC
Cooked, fried	3½ oz.	239	8.6	18.6	HC
Frozen, solids and liquid	3½ oz.	–	6.1	–	LP
Raw meat only	3½ oz.	66	8.4	3.4	LC
Pantry Pride Tuna in Oil	3½ oz.	243	24	–	HC
Perch, white	3½ oz.	118	19.3	0	HP
Perch, yellow	3½ oz.	91	19.5	0	HP
Pickerel, chain	3½ oz.	84	18.7	0	HP

Fish and Seafood

Food	Quantity	Calories	Protein Grams	Carbo-hydrates Grams	P/C/C Com-puter
Pike:					
Blue	3½ oz.	90	19.1	0	HP
Northern	3½ oz.	88	18.3	0	HP
Walleye	3½ oz.	93	19.3	0	HP
Pollock, cooked, creamed	3½ oz.	128	13.9	4.0	HC
Pompano	3½ oz.	166	18.8	0	HC
Red and gray snapper	3½ oz.	93	19.8	0	HP
Redhorse, silver	3½ oz.	98	18.0	0	HP
Redi-Jel Gefilte Fish	3½ oz.	52.6	7.74	1.57	LP
Rockfish, cooked, oven-steamed	3½ oz.	107	18.1	1.9	HP
Roe, cooked, baked or broiled	3½ oz.	126	22.0	1.9	HP
Roe, canned, solids and liquids	3½ oz.	118	21.5	.3	HP
S&W Nutradiet Blue-Green Label Salmon	3½ oz.	84	15.05	2.36	HP
Sablefish	3½ oz.	190	13.0	0	HC
Salmon (types and methods of preparing):					
Atlantic, canned	3½ oz.	203	21.7	0	HP
Atlantic, fresh	3½ oz.	217	22.5	0	HP
Chinook, canned	3½ oz.	210	19.6	0	HP
Chinook, fresh	3½ oz.	222	19.1	0	HP
Chum, canned	3½ oz.	139	21.5	0	HP
Chum, fresh	3½ oz.	–	–	–	–
Coho, canned	3½ oz.	153	20.8	0	HP
Coho, fresh	3½ oz.	–	–	–	–
Pink (humpback) canned	3½ oz.	141	20.5	0	HP
Pink (humpback) fresh	3½ oz.	119	20.0	0	HP
Sockeye (red) canned	3½ oz.	171	20.3	0	HP
Sockeye (red) fresh	3½ oz.	–	–	–	–
Methods of preparing:					
Cooked, broiled, or baked	3½ oz.	182	27.0	0	HP
Rice loaf	3½ oz.	122	12.0	7.3	LP
Smoked	3½ oz.	176	21.6	0	HP
Sardines, Atlantic, canned in oil	3½ oz.	311	20.6	.6	HC
Sardines, Pacific:					
Canned in brine or mustard	3½ oz.	196	18.8	1.7	HC
Canned in oil	3½ oz.	205	18.9	1.8	HC
Canned in tomato sauce	3½ oz.	197	18.7	1.7	HC
Sau-Sea Cooked Shrimp	3½ oz.	63	15.05	–	HP
Sau-Sea Shrimp Cocktail	3½ oz.	93.73	5.21	18.02	LP
Sauger	3½ oz.	84	17.9	0	LP

Fish and Seafood

Food	Quantity	Calories	Protein Grams	Carbo-hydrates Grams	P/C/C Com-puter
Scallops, bay and sea:					
Cooked, steamed	3½ oz.	112	23.2	–	HP
Frozen, breaded,					
fried, reheated	3½ oz.	194	18.0	10.5	HP
Sea bass, white	3½ oz.	96	21.4	0	HP
Shad or American shad	3½ oz.	201	23.2	0	HP
Shad, gizzard	3½ oz.	200	17.2	0	HP
Sheepshead, Atlantic, raw	3½ oz.	113	20.6	0	HP
Shrimp:					
Canned	3½ oz.	80	16.2	.8	HP
French-fried	3½ oz.	225	20.3	10.0	HC
Frozen, breaded	3½ oz.	139	12.3	19.9	HC
Shrimp or lobster paste,					
canned	3½ oz.	180	20.8	1.5	HP
Skate (raja fish)	3½ oz.	98	21.5	0	HP
Smelt, Atlantic, jack, and					
bay, fresh	3½ oz.	98	18.6	0	HP
Smelt, Atlantic, jack, and					
bay, canned	3½ oz.	200	18.4	0	HC
Snails	3½ oz.	90	6.1	2.0	LP
Spanish mackerel	3½ oz.	177	19.5	0	HP
Squid, fresh	3½ oz.	84	16.4	1.5	HP
Sturgeon, fresh	3½ oz.	94	18.1	0	HP
Sturgeon, smoked	3½ oz.	149	31.2	0	HP
Swanson Dinner Fillet of Ocean					
Fish	11½ oz.	440	27	38	HC
Swanson Fish 'n Chips	10¼ oz.	450	26	40	HC
Swordfish, fresh, cooked,					
broiled	3½ oz.	174	28.0	0	HC
Swordfish, canned, solids and					
liquids	3½ oz.	102	17.5	0	HP
Tautog (black fish)	3½ oz.	89	18.6	0	HP
Terrapin (diamond back) fresh	3½ oz.	111	18.6	0	HP
Tilefish, cooked, baked	3½ oz.	138	24.5	0	HP
Tomcod, Atlantic, fresh	3½ oz.	77	17.2	0	HP
Trout, brook, fresh	3½ oz.	101	19.2	0	HP
Trout, rainbow or steelhead,					
canned	3½ oz.	209	20.6	0	HP
Trout, rainbow or steelhead,					
fresh	3½ oz.	195	21.5	0	HP
Tuna:					
Bluefin, canned	3½ oz.	288	24.2	0	HC
Bluefin, fresh	3½ oz.	145	25.2	0	HP
Yellowfin, fresh	3½ oz.	133	24.7	0	HP
Turtle, green, canned	3½ oz.	106	23.4	0	HP
Turtle, green, fresh	3½ oz.	89	19.8	0	HP

Food	Quantity	Calories	Protein Grams	Carbo-hydrates Grams	P/C/C Com-puter
Underwood Sardines in Mustard Sauce	3½ oz.	197	14.9	2.2	HC
Underwood Sardines in Soya Bean Oil	3½ oz.	233	20.8	0.2	HC
Underwood Sardines in Tomato Sauce	3½ oz.	169	14.9	4.1	HC
Van Camp Sea Food Dietetic Tuna	3½ oz.	107	25	0.54	LCR
Van Camp Sea Food Light Chunk Tuna in Oil	3½ oz.	253.08	24.23	0.54	HC
Van Camp Sea Food White Solid Tuna in Oil	3½ oz.	285	25	0.5	HC
Van Camp Sea Food White Solid Tuna in Water	3½ oz.	170	25	0.5	LC
Weakfish, cooked, broiled	3½ oz.	208	24.6	0	HC
Weight Watchers Fille 'o' Fish Dinner	18 oz.	266	7.9	4.3	HC
Weight Watchers Fille 'o' Fish Luncheon	9½ oz.	175	8.1	6.4	HC
Weight Watchers Flounder Dinner	16 oz.	268	7.4	3.84	HC
Weight Watchers Flounder Luncheon	9½ oz.	170	6.7	3.27	HC
Weight Watchers Greenland Turbot Dinner	18 oz.	426	6.4	3.4	HC
Weight Watchers Greenland Turbot Luncheon	9½ oz.	277	7.9	4.1	HC
Weight Watchers Ocean Perch Dinner	18 oz.	307	8.7	3.6	HC
Weight Watchers Ocean Perch Luncheon	9½ oz.	185	7.5	2.65	HC
Weight Watchers Sole Dinner	18 oz.	279	8.6	3.6	HC
Weight Watchers Sole Luncheon	9½ oz.	190	8.6	3	HC
Whale meat, fresh	3½ oz.	156	20.6	0	HC
Whitefish, lake, cooked, baked, stuffed	3½ oz.	215	15.2	5.8	HC
Whitefish, lake, smoked	3½ oz.	155	20.9	0	HC
Wreckfish, fresh	3½ oz.	114	18.4	0	HP
Yellowtail, fresh	3½ oz.	138	21.0	0	HP

Fish and Seafood

Dairy Products, Eggs

Food	Quantity	Calories	Protein Grams	Carbo- hydrates Grams	P/C/C Com- puter
American pasteurized process cheese	3½ oz.	370	23.2	1.9	HC
American pasteurized process cheese food	3½ oz.	323	19.8	7.1	HC
American pasteurized process cheese spread	3½ oz.	288	16.0	8.2	HC
Banquet Cooking Bag Macaroni and Cheese	8 oz.	280	44	144	HCR
Banquet Dinner Macaroni and Cheese	12 oz.	326	13.3	45.6	HC
Banquet Entrée Macaroni and Cheese	8 oz.	279	11.8	35.9	HC
Blue or Roquefort cheese	3½ oz.	368	21.5	2.0	HC
Bosco Milk Amplifier	1 tbsp.	50	–	12	LC
Brick cheese	3½ oz.	370	22.2	1.9	HC
Buttermilk (from skim milk)	3½ oz.	36	3.6	5.1	LCR
Camembert cheese (domestic)	3½ oz.	299	17.5	1.8	HC
Celeste Cheese Dinner Ravioli	7.5 oz.	255	11.5	38.3	HC
Celeste Cheese Pizza	5 oz.	321	16.4	35.6	HC
Celeste Cheese Ravioli	4 oz.	264	13.4	38.3	HC
Celeste Manicotti	13 oz.	428	22.4	40.6	HC
Cheddar (domestic, called American)	3½ oz.	398	25.0	2.1	HC
Cheese fondue from home recipe	3½ oz.	265	14.8	10.0	HC
Cheese soufflé from home recipe	3½ oz.	218	9.9	6.2	HC
Cheese straws	3½ oz.	453	11.2	34.5	HC
Chiffon Soft Diet Margarine	1 tbsp.	50	0	0	LC
Chiffon Soft Margarine	1 tbsp.	100	0	0	HC
Chiffon Unsalted Margarine	1 tbsp.	100	0	0	HC
Chiffon Whipped Soft Margarine	1 tbsp.	70	0	0	LC
Chocolate milk	3½ oz.	520	7.7	56.9	HC
Cottage cheese, creamed	3½ oz.	106	13.6	2.9	HC
Cottage cheese, uncreamed	3½ oz.	86	17.0	2.7	LC
Cream cheese	3½ oz.	374	8.0	2.1	HC
Cream, fluid:					
Half-and-half (cream and milk)	3½ oz.	134	3.2	4.6	HC
Heavy whipping	3½ oz.	352	2.2	3.1	HC
Light, coffee, or table	3½ oz.	211	3.0	4.3	HC
Light whipping	3½ oz.	300	2.5	3.6	HC
Cream substitute, dried	3½ oz.	508	8.5	61.3	HC
Dannon Apricot Yogurt	8 oz.	260	7.8	51.5	HC

Food	Quantity	Calories	Protein Grams	Carbo-hydrates Grams	P/C/C Computer
Dannon Blueberry Yogurt	8 oz.	260	7.8	51.5	HC
Dannon Boysenberry Yogurt	8 oz.	260	7.8	51.5	HC
Dannon Cherry Yogurt	8 oz.	260	7.8	51.5	HC
Dannon Coffee Yogurt	8 oz.	200	9.2	33.6	HC
Dannon Dutch Apple Yogurt	8 oz.	260	7.8	51.5	HC
Dannon Peach Yogurt	8 oz.	260	7.8	51.5	HC
Dannon Pineapple-Orange Yogurt	8 oz.	260	7.8	51.5	HC
Dannon Plain Yogurt	8 oz.	130	11	14.2	LC
Dannon Prune Whip Yogurt	8 oz.	260	7.8	51.5	HC
Dannon Red Raspberry Yogurt	8 oz.	260	7.8	51.5	HC
Dannon Strawberry Yogurt	8 oz.	260	7.8	51.5	HC
Dannon Vanilla Yogurt	8 oz.	200	9.2	33.6	HC
Dorman's Bonbel Cheese	1 oz.	97	6	51	HCR
Dorman's Fondine Cheese	1 oz.	85	3.2	–	LC
Dorman's Laughing Cow Cheese	1 oz.	80	5	1	LC
Dorman's Mozzarella	1 oz.	77	9	54	HCR
Dorman's Muenster Cheese	1 oz.	109	7	16	HC
Dorman's Natural Swiss Cheese	1 oz.	108	8	27	LP
Dorman's Port Salut Cheese	1 oz.	98.8	7	.08	LP
Dorman's Tilsit Cheese	1 oz.	96.7	7	19	LP
Egg White	2 oz.	30	8	–	LC
Egg Yolk	2 oz.	120	6	–	HC
Eggs:					
Fried	3½ oz.	216	13.8	.3	HC
Hard-Cooked	3½ oz.	163	12.9	.9	HC
Omelet	3½ oz.	173	11.2	2.4	HC
Poached	3½ oz.	163	12.7	.8	HC
Scrambled	3½ oz.	173	11.2	2.4	HC
Fisher's Cheez-ola	1 oz.	89.71	7.08	1.13	LC
Fisher's Count-Down Cheese	1 oz.	41.88	6.79	2.83	LC
Fisher's Spreadable Count-Down Cheese	1 oz.	34.24	5.66	2.26	LC
Ice cream and frozen custard:					
Regular with 10% fat	3½ oz.	193	4.5	20.8	HC
Regular with 12% fat	3½ oz.	207	4.0	20.6	HC
Rich with 16% fat	3½ oz.	222	2.6	18.0	HC
Ice milk	3½ oz.	152	4.8	22.4	HC
Jeno's Cheese Pizza	7.5 oz.	550	17	80	HC
La Choy Skillet Dinners					
Egg Foo Young	3½ oz.	112	3.96	5.98	LP
Land 'o' Lakes Buttermilk	8 oz.	95	9	12	LC
Land 'o' Lakes Chocolate-Flavored Skim Milk	8 oz.	150	9	26	HC
Land 'o' Lakes Chocolate Milk	8 oz.	210	9	26	HC

Food	Quantity	Calories	Protein Grams	Carbo- hydrates Grams	P/C/C Com- puter
Land 'o' Lakes Eggs:	2 small	125	10	1	HP
	2 medium	150	12	1	HP
	2 large	165	13	1	HP
	2 extra-large	180	14	1	HP
	2 jumbo	210	17	1	HP
Land 'o' Lakes Golden Velvet Cheese	1 oz.	80	5	2	LC
Land 'o' Lakes Low-Fat Chocolate-Flavored Milk	8 oz.	190	9	27	HC
Land 'o' Lakes Natural Cheddar	1 oz.	115	7	1	HC
Land 'o' Lakes Natural Colby	1 oz.	110	6	1	HC
Land 'o' Lakes Natural Swiss Cheese	1 oz.	110	8	1	HC
Land 'o' Lakes Processed American Cheese	1 oz.	115	7	1	HC
Land 'o' Lakes Regular Margarine	1 tbsp.	100	0	0	HC
Land 'o' Lakes Skim Milk	8 oz.	95	9	12	LC
Land 'o' Lakes Slim and Light Milk	8 oz.	115	10	14	HP
Land 'o' Lakes Soft Margarine	1 tbsp.	90	0	0	LC
Land 'o' Lakes 2% Fat Milk	8 oz.	130	9	12	HP
Land 'o' Lakes Whole Milk	8 oz.	160	9	12	HC
Limburger cheese	3½ oz.	345	21.2	2.2	HC
Lucerne Evaporated Milk	8 oz.	340	18	24	HC
Mazola Diet Imitation Margarine	1 tbsp.	50	0	0	LC
Mazola Regular Margarine	1 tbsp.	100	0	0	LC
Milk:					
Condensed (sweetened)	3½ oz.	321	8.1	54.3	HC
Dry, whole	3½ oz.	502	26.4	38.2	HC
Dry, skim (nonfat solids) regular	3½ oz.	363	35.9	52.3	HC
Dry, skim (nonfat solids) instant	3½ oz.	359	35.8	51.6	HC
Evaporated (unsweetened)	3½ oz.	137	7.0	9.7	HC
Goat	3½ oz.	67	3.2	4.6	LC
Half-and-half (cream and milk)	3½ oz.	134	3.2	4.6	HC
Partially skimmed with 2% nonfat milk solids	3½ oz.	59	4.2	6.0	LC
Skim	3½ oz.	36	3.6	5.1	LC
Whole	3½ oz.	65	3.5	4.9	LC

Food	Quantity	Calories	Protein Grams	Carbo-hydrates Grams	P/C/C Computer
Morton Casserole Macaroni and Cheese	8 oz.	296.54	13.35	31.52	HC
Morton Macaroni and Cheese	12.75 oz.	316.62	12.64	38.53	HC
Nucoa Regular Margarine	1 tbsp.	100	0	0	LC
Nucoa Soft Margarine	1 tbsp.	90	0	0	LC
Oregon Freeze Dry Cheese Omelette	3½ oz.	490	41	23	HC
Oregon Freeze Dry Eggs w/Bacon Bits	3½ oz.	510	42	21	HC
Oregon Freeze Dry Mexican Omelette	3½ oz.	480	38	23	HC
Oregon Freeze Dry Scrambled Eggs w/Butter	3½ oz.	490	40	25	HC
Ovaltine, Chocolate Flavor	1 oz.	111	1.4	24	HC
Ovaltine, Malt Flavor	1 oz.	110	2.2	22.9	HC
Pantry Pride Dry nonfat Milk	8 oz.	80	8	12	HP
Pantry Pride Evaporated Milk	8 oz.	350	18	24	HC
Parmesan cheese	3½ oz.	393	36.0	2.9	HC
Pimiento pasteurized process cheese	3½ oz.	371	23.0	1.8	HC
Safeway Regular, Corn, Tub Margarine	1 tbsp.	100	0	0	LC
Saffola Cube Margarine	3½ oz.	720	.06	.04	HC
Saffola Soft Margarine	3½ oz.	720	0.6	0.4	HC
Sanalac Imitation Instant Milk	8 oz.	80	8	12	HP
Swiss (domestic) cheese	3½ oz.	370	27.5	1.7	HC
Swiss pasteurized process cheese	3½ oz.	355	26.4	1.6	HC
Snack Mate American Cheese Spread	1 oz.	82.64	4.47	2.26	LC
Snack Mate Cheddar Cheese Spread	1 oz.	83.2	4.53	2.26	LC
Snack Mate French Onion Cheese Spread	1 oz.	82.35	4.39	2.41	LC
Snack Mate Hickory Smoke Cheese Spread	1 oz.	82.07	4.36	2.69	LC
Snack Mate Pimiento Cheese Spread	1 oz.	83.7	4.53	2.18	LC
Snack Mate Seasoned Cheddar Cheese Spread	1 oz.	82.64	4.47	2.69	LC
Soybean milk, fluid	3½ oz.	33	3.4	2.2	LC
Soybean milk, powder	3½ oz.	429	41.8	28.0	HC
Soybean milk products, sweetened:					
Liquid concentrate	3½ oz.	126	4.8	12.3	HC
Powder	3½ oz.	452	20.4	48.4	HC

Food	Quantity	Calories	Protein Grams	Carbo-hydrates Grams	P/C/C Com-puter
Welsh rarebit	3½ oz.	179	8.1	6.3	HC
Whey, dried	3½ oz.	349	12.9	73.5	HC
Whey, fluid	3½ oz.	26	.9	5.1	LC
Yogurt from partially skimmed milk	3½ oz.	50	3.4	5.2	LC
Yogurt from whole milk	3½ oz.	62	3.0	4.9	LC

Dairy Products, Eggs

Beans, Nuts, Seeds

Food	Quantity	Calories	Protein Grams	Carbo-hydrates Grams	P/C/C Com-puter
Almonds, dried	3½ oz.	598	18.6	19.5	HC
Almonds, roasted and salted	3½ oz.	627	18.6	19.5	HC
Almonds, sugar-coated	3½ oz.	456	7.8	70.2	HC
Armour-Dial Chili w/Beans	7¾ oz.	370	15	26	HC
Armour-Dial Chili w/o Beans	7½ oz.	430	13	15	HC
B & M Baked Beans	3½ oz.	150	6.9	22.7	HCR
B & M Baked Red Kidney Beans	3½ oz.	158	8.8	21.8	HCR
B & M Baked Yellow Eye Beans	3½ oz.	159	6.8	22.4	HCR
B & M Shelled Beans	3½ oz.	125	9.1	20.9	HCR
Beans:					
Common, white	3½ oz.	118	7.8	21.2	HCR
Common, white, canned with pork-sweet sauce	3½ oz.	150	6.2	21.1	HCR
Common, white, canned with pork-tomato sauce	3½ oz.	122	6.1	19.0	HCR
Common, white, canned without pork	3½ oz.	120	6.3	23.0	HCR
Lima, canned, drained	3½ oz.	96	5.4	18.3	LC
Lima, cooked, drained	3½ oz.	111	7.6	19.8	HCR
Lima, frozen	3½ oz.	99	6.0	19.1	HCR
Lima, low-sodium dietary pack, cooked, drained	3½ oz.	95	5.8	17.7	LC
Mung, cooked, drained	3½ oz.	28	3.2	5.2	LC
Red, canned	3½ oz.	90	5.7	16.4	LC
Red, cooked	3½ oz.	118	7.8	21.4	HCR
Snap, canned, drained	3½ oz.	24	1.4	5.2	LCR
Snap, cooked, drained	3½ oz.	25	1.6	5.4	LCR
Snap, frozen, cooked, drained	3½ oz.	25	1.6	5.7	LCR
Snap, low-sodium dietary pack, cooked, drained	3½ oz.	22	1.5	4.8	LCR
Yellow or wax, canned, drained	3½ oz.	24	1.4	5.2	LCR
Yellow or wax, cooked, drained	3½ oz.	22	1.4	4.6	LCR
Yellow or wax, frozen, cooked, drained	3½ oz.	27	1.7	6.2	LCR
Yellow or wax, low-sodium dietary pack, cooked, drained	3½ oz.	21	1.2	4.7	LCR
Beechnuts	3½ oz.	568	19.4	20.3	HC

Food	Quantity	Calories	Protein Grams	Carbo-hydrates Grams	P/C/C Com-puter
Brazil nuts	3½ oz.	654	14.3	10.9	HC
Broadbeans	3½ oz.	338	25.1	58.2	HC
Broadcast Chili w/Beans	3½ oz.	170	6.7	10.9	HC
Broadcast Chili w/o Beans	3½ oz.	210	5.94	5.2	HC
Bulgur	3½ oz.	359	8.7	79.5	HC
Butternuts	3½ oz.	629	23.7	8.4	HC
Campbell's Barbecue Beans	3½ oz.	122.5	5	20	HC
Campbell's Beans and Pork w/Tomato Sauce	3½ oz.	114	5	19	HC
Campbell's Beans in Molasses w/Tomato Sauce	3½ oz.	127	5	21	HC
Campbell's Home Style Beans	3½ oz.	131	5	23	HC
Cashew nuts	3½ oz.	561	17.2	29.3	HC
Chestnuts, dried	3½ oz.	377	6.7	78.6	HC
Chestnuts, fresh	3½ oz.	194	2.9	42.1	HCR
Chick-peas (garbanzos)	3½ oz.	360	20.5	61.0	HC
Cowpeas, canned, boiled, drained	3½ oz.	70	5.0	12.4	LC
Cowpeas, cooked, boiled, drained	3½ oz.	108	8.1	18.1	HC
Del Monte Green Beans	½ cup	15.5	0.85	2.8	LC
Del Monte Italian Beans	½ cup	24.5	1.4	4.65	LC
Del Monte Lima Beans	½ cup	88.5	5.15	16.45	LC
Del Monte Seasoned Green Beans	½ cup	15	0.8	2.55	LC
Del Monte Seasoned Lima Beans	½ cup	88	6.2	15.85	LC
Del Monte Wax Beans	½ cup	14	0.55	2.45	LC
Filberts (hazelnuts)	3½ oz.	634	12.6	16.7	HC
Green Giant Green Beans	3½ oz.	13	0.7	2.3	LP
Green Giant LeSueur Three-Bean Salad	3½ oz.	65	2.0	12.0	LP
Green Giant Wax Beans	3½ oz.	12	0.8	2.2	LP
Heinz Chili w/Beans	3½ oz.	169	7.8	11.7	HC
Heinz Single Serving Beans and Pork w/Tomato Sauce	3½ oz.	111	6.1	18.3	HC
Heinz Single Serving Beans in Molasses w/Tomato Sauce	3½ oz.	108	6.1	20.6	HC
Heinz Single Serving Beans in Tomato Sauce	3½ oz.	103	6.0	18.5	HC
Hickory nuts	3½ oz.	673	13.2	12.8	HC
Homemaker's Baked Beans	3½ oz.	141	6.5	21.5	HC
Homemaker's Baked Red Kidney Beans	3½ oz.	149	6.8	22.2	HC
Hunt's Beans and Pork w/Tomato Sauce	3½ oz.	116	5.6	22.2	HC
Hunt's Canned Chili Beans	3½ oz.	117	7.2	20.8	HC
Hunt's Kidney Beans	½ cup	118	7.1	22.2	HC

Food	Quantity	Calories	Protein Grams	Carbo-hydrates Grams	P/C/C Com-puter
Hunt's Small Red Beans	½ cup	110	6.8	20.4	HC
Hunt's Snack Pack Beans and Pork w/Tomato Sauce	5 oz.	169	8.0	33.8	HC
Hunt's Snack Pack Three-Bean Salad	5 oz.	111	6.7	22.2	HC
Hyacinth beans	3½ oz.	338	22.2	61.0	HC
Lentils, cooked	3½ oz.	106	7.8	19.3	HC
Loquats	3½ oz.	48	.4	12.4	LC
Lychees, dried	3½ oz.	277	3.8	70.7	HC
Macadamia nuts	3½ oz.	691	7.8	15.9	HC
Nu Made Peanut Butter	2 tbsp.	200	8	6	HC
Oregon Freeze Dry Chili Mac	3½ oz.	440	24.5	54	HC
Oregon Freeze Dry Chili w/Beans	3½ oz.	430	44	31	HC
Pantry Pride Green Beans	½ cup	20	1	4	LC
Peanuts:					
Boiled	3½ oz.	376	15.5	14.5	HC
Raw, with skins	3½ oz.	564	26.0	18.6	HC
Raw, without skins	3½ oz.	568	26.0	17.6	HC
Roasted and salted	3½ oz.	585	26.0	18.8	HC
Roasted, with skins	3½ oz.	582	26.2	20.6	HC
Pecans	3½ oz.	687	9.2	14.6	HC
Peter Pan Smooth Peanut Butter	2 tbsp.	183	13	6	HC
Pignolia nuts	3½ oz.	552	31.1	11.6	HC
Pili nuts	3½ oz.	669	11.4	8.4	HC
Piñon nuts	3½ oz.	635	13.0	20.5	HC
Pistachio nuts	3½ oz.	594	19.3	19.0	HC
Pumpkin and squash seed kernels, dry	3½ oz.	553	29.0	15.0	HC
Royal Hawaiian Macadamia Nuts	¼ cup	394	4.4	9.06	HC
Safflower seed kernels, dry	3½ oz.	615	19.1	12.4	HC
Sesame seeds, dry	3½ oz.	563	18.6	21.6	HC
Skippy Dry Roasted Cashews	1 oz.	160	6	8	HC
Skippy Dry Roasted Mixed Nuts	1 oz.	170	6	5	HC
Skippy Dry Roasted Peanuts	1 oz.	160	9	4	HC
Skippy Peanut Butter	2 tbsp.	190	9	4	HC
Smucker's Grape Goober Peanut Butter	2 tbsp.	120	3.5	14	HC
Smucker's Peanut Butter	2 tbsp.	170	8	6	HC
Soybean curd (tofu)	3½ oz.	72	7.8	2.4	LC
Soybean protein	3½ oz.	322	74.9	15.1	HC
Soybean proteinate	3½ oz.	312	80.6	7.7	HC
Soybeans, canned, solids and liquid	3½ oz.	75	6.5	6.3	LC
Soybeans, cooked, boiled, drained	3½ oz.	118	9.8	10.1	HP
Stokely-Van Camp Green Beans	½ cup	20	1.1	4.6	LP

Beans, Nuts, Seeds

Food	Quantity	Calories	Protein Grams	Carbo-hydrates Grams	P/C/C Com-puter
Stokely-Van Camp Lima Beans	½ cup	82	4.7	15.4	LP
Stokely-Van Camp Wax Beans	½ cup	22	1.1	4.6	LP
Sunflower seed kernels	3½ oz.	560	24.0	19.9	HC
Town House Green Beans	½ cup	20	1	4	LP
Van Camp Beans and Pork w/Tomato Sauce	½ cup	143	7	20.9	HC
Van Camp Beans in Tomato Sauce	½ cup	138	6.9	26	HC
Van Camp Chili w/Beans	1 cup	304	17.2	28	HC
Van Camp Chili w/o Beans	1 cup	460	23.6	13.2	HC
Walnuts, Black	3½ oz.	628	20.5	14.8	HC
Walnuts, Persian or English	3½ oz.	651	14.8	15.8	HC
Water chestnuts (Chinese), fresh	3½ oz.	79	1.4	19.0	LP

Fats and Oils

Food	Quantity	Calories	Protein Grams	Carbo-hydrates Grams	P/C/C Com-puter
Butter	3½ oz.	716	.6	.4	HC
Regular	1 tbsp.	100	–	–	HC
Whipped	1 tbsp.	65	–	–	LC
Oil (or dehydrated butter)	3½ oz.	876	.3	0	HC
Fats, cooking (vegetable fat)	3½ oz.	884	0	0	HC
Lard	3½ oz.	902	0	0	HC
Margarine	3½ oz.	720	.6	.4	HC
Oils, salad or cooking	3½ oz.	884	0	0	HC
Salad dressings, commercial:					
Blue and Roquefort (cheese)	3½ oz.	504	4.8	7.4	HC
French	3½ oz.	410	.6	17.5	HC
Italian	3½ oz.	552	.2	6.9	HC
Mayonnaise	3½ oz.	718	1.1	2.2	HC
Russian	3½ oz.	494	1.6	10.4	HC
Thousand Island	3½ oz.	502	.8	15.4	HC
Salad dressings, home recipe:					
French	3½ oz.	632	.3	3.6	HC
Cooked	3½ oz.	164	4.4	15.2	HC
Suet (beef kidney fat)	3½ oz.	854	1.5	0	HC

Condiments, Sauces

Food	Quantity	Calories	Protein Grams	Carbo-hydrates Grams	P/C/C Com-puter
B&M Corn Relish	1 tbsp.	27.45	.453	6.39	LP
Barbecue sauce	3½ oz.	91	1.5	8.0	LC
Celeste Italian Sauce	4 oz.	68	2.3	9.2	LC
Del Monte Catsup	1 tbsp.	30.2	0.34	7.84	LC
Del Monte Chili Sauce	1 oz.	32.88	0.62	8.49	LC
Del Monte Dill Pickles	1 large	7	0.4	1.4	LC
Del Monte Fresh Pickles	1 medium	18	0.9	3.9	LC
Del Monte Hamburger Pickle Relish	1 tbsp.	37.07	0.17	10.05	LC
Del Monte Hot Dog Pickle Relish	1 tbsp.	31.7	0.54	7.84	LC
Del Monte Seafood Cocktail Sauce	½ cup	159.5	2.25	42	HC
Del Monte Sour Pickles	1 large	10	0.3	2	LC
Del Monte Sweet Pickle Relish	1 tbsp.	40.47	.17	10.47	HCR
Del Monte Tomato Paste	½ cup	108.5	4.9	25.05	HCR
Del Monte Tomato Sauce	½ cup	30	1.75	6.35	LC
Del Monte Tomato w/Mushrooms Sauce	½ cup	37.5	1.9	8.5	LC
Del Monte Tomato w/Onions Sauce	½ cup	45	2	9.25	LC
Del Monte Tomato w/Tomato Tidbits Sauce	½ cup	45	1.4	10.9	LC
Dromedary Pimientos, Pieces	1 tbsp.	7.64	0.25	1.16	LC
Dromedary Sliced Pimientos	1 tbsp.	7.64	0.25	1.16	LC
Dromedary Whole Pod Pimientos	1 tbsp.	7.64	0.25	1.16	LC
Durkee a la King Sauce	½ cup	136.5	1.5	12.75	HC
Durkee au jus Gravy Mix	¼ cup	17.5	0.7	3.7	LC
Durkee Beef Stew Packaged Mix Seasonings	1 pkg.	174	4.5	38.1	HC
Durkee Brown Gravy Mix	¼ cup	24	0.57	2.8	LC
Durkee Cheese Sauce	½ cup	261.5	14	15.5	HC
Durkee Chicken Dip'n Seasoning Coating	1 pkg.	286	3.9	27.8	HC
Durkee Chicken Gravy Mix	¼ cup	24	0.57	2.8	LC
Durkee Chili con Carne Sauce	½ cup	78.4	1.9	17.84	HCR
Durkee Chop Suey Sauce	1 pkg.	128	1.8	18.6	HC
Durkee Famous Sauce	1 tbsp.	69	0.55	2.21	LC
Durkee Ground Beef Packaged Mix Seasonings	1 pkg.	94	2.4	19.9	LC
Durkee Hollandaise Sauce	½ cup	159.75	8.25	8.25	HC
Durkee Imitation Bacon Bits	1 tsp.	8	0.07	0.05	LC
Durkee Imitation Butter Flavor Salt	1 tsp.	3	0.15	0.02	LC
Durkee Lemon Pepper	1 tsp.	1	.004	.02	LC
Durkee Mushroom Gravy Mix	¼ cup	18.3	0.4	3	LC

Food	Quantity	Calories	Protein Grams	Carbo-hydrates Grams	P/C/C Com-puter
Durkee Onion Gravy Mix	¼ cup	24	0.27	4	LC
Durkee Salad Mate	1 tsp.	7	0.05	1	LC
Durkee Salad Seasoning	1 tsp.	4	0.02	0.07	LC
Durkee Salad Seasoning w/Cheese	1 tsp.	10	0.05	0.04	LC
Durkee Sloppy Joe Sauce	½ cup	38	–	9.35	LC
Durkee Sour Cream Sauce	½ cup	243.75	14.25	22.5	HC
Durkee Spaghetti Sauce	½ cup	41.75	1.5	9.7	LC
Durkee Spaghetti w/Mushrooms Sauce	½ cup	39.19	2.06	9.02	LC
Durkee Sweet-Sour Sauce	½ cup	115	0.55	22.3	HC
Durkee White Sauce	½ cup	190.5	5	12.75	HC
Fanning's Bread and Butter Pickles	3½ oz.	45	0.6	11.5	LC
Franco American Beef Gravy	2 oz.	35	2	3	LC
Franco American Brown Gravy w/Onions	2 oz.	35	1	4	LC
Franco American Chicken Gravy	2 oz.	55	1	3	LC
Franco American Chicken Giblet Gravy	2 oz.	35	1	3	LC
Franco American Mushroom Gravy	2 oz.	35	1	4	LC
French's au jus Gravy Mix	¼ cup	5	0.1	1.1	LC
French's Barbeque Seasoning	1 tsp.	7	0.03	0.07	LC
French's Beef Flavor Stock Base Seasoning	1 tsp.	9	0.04	1.17	LC
French's Brown Gravy Mix	¼ cup	18	0.6	2.9	LC
French's Brown and Spicy Mustard	1 tbsp.	17	0.9	1.4	LC
French's Celery Spice	1 tsp.	2	0.02	0.01	LC
French's Cheese Sauce	½ cup	162	8.6	11.6	HC
French's Chicken Flavor Stock Base	1 tsp.	8	0.03	1.2	LC
French's Chicken Gravy Mix	¼ cup	32	1.8	3.9	LC
French's Cinnamon Sugar	1 tsp.	15	–	3.8	LC
French's Cream-Style Mustard	1 tbsp.	11	0.6	0.9	LC
French's Garlic Salt	1 tsp.	4	0.01	0.07	LC
French's Garlic Salt, Parslied	1 tsp.	6	0.01	1.2	LC
French's Ground Beef w/Onions Packaged Mix Seasonings	1 pkg.	324	20.8	4.4	HC
French's Hickory Smoke Salt	1 tsp.	1	0.01	–	LC
French's Hollandaise Sauce	1 tbsp.	16	0.04	0.06	LC
French's Imitation Bacon Crumbies	1 tsp.	7	0.06	0.04	LC
French's Imitation Butter Flavor Salt	1 tsp.	8	–	–	LC

Food	Quantity	Calories	Protein Grams	Carbo-hydrates Grams	P/C/C Computer
French's Lemon and Pepper Seasoning	1 tsp.	5	–	1	LC
French's Meat Tenderizer	1 tsp.	2	.02	.02	LC
French's Meat Tenderizer, Seasoned	1 tsp.	2	.02	.01	LC
French's Medford Mustard	1 tbsp.	15	0.6	1.2	LC
French's Mild Barbeque Sauce	1 tbsp.	17	0.02	4.1	LC
French's Mushroom Gravy Mix	¼ cup	16	1.3	1.7	LC
French's Mustard w/Horseradish	1 tbsp.	17	0.7	1.2	LC
French's Mustard w/Onion	1 tbsp.	21	0.7	3.4	LC
French's Onion Gravy Mix	¼ cup	18	0.7	3.1	LC
French's Onion Salt	1 tsp.	5	.01	1	LC
French's Pizza Sauce	1 tbsp.	9	0.03	2.05	LC
French's Pizza Seasoning	1 tsp.	4	0.02	0.05	LC
French's Pork Gravy Mix	¼ cup	19	1.1	2.6	LC
French's Regular Barbeque Sauce	1 tbsp.	9	0.02	2.1	LC
French's Ring Star Mustard	1 tbsp.	12	0.6	1.2	LC
French's Salad Lift	1 tsp.	6	0.02	1.2	LC
French's Seafood Seasoning	1 tsp.	2	0.01	0.03	LC
French's Seasoned Pepper	1 tsp.	7	0.03	1	LC
French's Seasoning Salt	1 tsp.	3	–	0.07	LC
French's Sloppy Joe Packaged Season Mix	½ cup	184	11.2	7.2	HC
French's Smoky Barbeque Sauce	1 tbsp.	15	0.02	3.7	LC
French's Sour Cream Sauce	1 tbsp.	25	0.06	1.7	LC
French's Spaghetti Sauce	½ cup	89.6	1.76	10.72	HC
French's Spaghetti w/Mushrooms Sauce	½ cup	72.8	2.64	9.84	HC
French's Stroganoff Sauce	1/3 cup	104	4.4	9.3	HC
French's Taco Sauce	1/3 cup	223	14.3	4	HC
French's Turkey Gravy Mix	¼ cup	23	1.0	2.9	LC
French's Worcestershire Sauce	1 tbsp.	6	0.01	1.4	LC
Good Seasons Blue Cheese Salad Dressing Mix	1 tbsp.	89	0.01	1.3	HC
Good Seasons Cheese Garlic Salad Dressing Mix	1 tbsp.	84	0.01	0.08	HC
Good Seasons Cheese Italian Salad Dressing Mix	1 tbsp.	89	0.01	1.3	HC
Good Seasons French Riviera Salad Dressing Mix	1 tbsp.	90	0.01	2.4	HC
Good Seasons Garlic Salad Dressing Mix	1 tbsp.	84	0.01	0.08	HC
Good Seasons Italian Salad Dressing Mix	1 tbsp.	84	0.01	0.08	HC
Good Seasons Low-Calorie Italian Salad Dressing Mix	1 tbsp.	3	0.01	0.07	LC

Condiments, Sauces

Food	Quantity	Calories	Protein Grams	Carbo-hydrates Grams	P/C/C Computer
Good Seasons Mild Italian Salad Dressing Mix	1 tbsp.	89	0.01	1.3	HC
Good Seasons Old-Fashioned French Salad Dressing Mix	1 tbsp.	83	0.01	0.05	HC
Good Seasons Onion Salad Dressing Mix	1 tbsp.	84	0.01	0.08	HC
Good Seasons Thick'n Creamy Blue Cheese Salad Dressing Mix	1 tbsp.	96	0.03	0.09	HC
Good Seasons Thick'n Creamy French Salad Dressing Mix	1 tbsp.	97	0.02	1.9	HC
Good Seasons Thick'n Creamy Italian Salad Dressing Mix	1 tbsp.	94	0.02	1	HC
Good Seasons Thousand Island Salad Dressing Mix	1 tbsp.	80	0.02	1.7	HC
Hellmann's French Garlic Salad Dressing	1 tbsp.	70	–	3	HC
Hellmann's French Salad Dressing	1 tbsp.	60	–	3	HC
Hellmann's Mayonnaise	1 tbsp.	100	0	0	HC
Hellmann's Spin Blend Salad Dressing	1 tbsp.	60	–	3	HC
Hellmann's Tartar Sauce Salad Dressing	1 tbsp.	70	–	–	HC
Hunt's Catsup	1 tbsp.	18.2	0.29	5.1	LC
Hunt's Distilled Vinegar	3.5 oz.	12	–	5	LC
Hunt's Special Tomato Sauce	½ cup	47.1	1.65	10.98	HCR
Hunt's Tomato Herb Sauce	½ cup	93	2.1	13.04	HC
Hunt's Tomato Paste	1 tbsp.	13.5	0.56	0.05	LC
Hunt's Tomato Sauce	½ cup	39.25	1.7	8.95	HC
Hunt's Tomato w/Bits Sauce	½ cup	41	1.75	9.3	HC
Hunt's Tomato w/Cheese Sauce	½ cup	53.5	2.5	9.95	HC
Hunt's Tomato w/Mushrooms Sauce	½ cup	41.6	1.75	9.45	HC
Hunt's Tomato w/Onions Sauce	½ cup	51	2	11.65	HC
La Choy Brown Sauce	½ cup	642.6	3.4	157	HCR
La Choy Soy Sauce	½ cup	101.7	17.2	8.3	HC
La Choy Sweet-Sour Sauce	½ cup	244	0.03	54.7	HC
Lawry's au jus Gravy Mix	1 oz.	8.75	2.08	3.63	LC
Lawry's BacOnion Seasoning	1 oz.	106.29	1.75	5.76	HC
Lawry's Bacon Salad Dressing Mix	1 pkg.	69	14	53	HC
Lawry's Barbeque Sweet and Sour Sauce	1 oz.	36.93	1.79	4.05	LC

Food	Quantity	Calories	Protein Grams	Carbo-hydrates Grams	P/C/C Computer
Lawry's Beef Marinade Packaged Mix Seasonings	1 pkg.	69	3.85	3.27	HC
Lawry's Beef Olé Packaged Mix Seasonings	1 pkg.	125.8	1.13	6.17	HC
Lawry's Beef Stew Packaged Mix Seasonings	1 pkg.	131.1	1.21	5.20	HC
Lawry's Blue Cheese Salad Dressing Mix	1 pkg.	79	0.02	0.1	HC
Lawry's Brown Gravy Mix	1½ oz.	161	16.4	45.3	HC
Lawry's Burgundy Wine Sauce	1 pkg.	98.1	1.56	5.75	HC
Lawry's California Ranch Salad Dressing Mix	1 pkg.	84.51	0.03	0.07	HC
Lawry's Canadian Salad Dressing	2 tbsp.	143.4	2.8	4.2	HC
Lawry's Caesar Salad Dressing Mix	1 pkg.	71	0.06	0.09	HC
Lawry's Champagne-Style Salad Dressing	2 tbsp.	128.3	–	10.4	HC
Lawry's Cheese Italian Salad Dressing Mix	1 pkg.	68.9	0.04	0.07	HC
Lawry's Chicken Gravy Mix	1½ oz.	164	15.6	48	HC
Lawry's Chili Packaged Mix Seasoning	1 pkg.	136.6	1.24	5.4	HC
Lawry's Chili Powder	1 oz.	3	0.04	0.06	LC
Lawry's Enchilada Sauce	1 pkg.	144	9.63	5.9	HC
Lawry's Family-Style Spaghetti Sauce	1 pkg.	208.9	9.27	5.63	HC
Lawry's French Salad Dressing	2 tbsp.	119.45	1	12	HC
Lawry's Garlic Salt	1 oz.	40.46	0.02	0.01	LC
Lawry's Goulash Packaged Mix Seasoning	1 pkg.	126.9	1.16	5.28	HC
Lawry's Green Goddess Salad Dressing	2 tbsp.	118.5	2.3	4.8	HC
Lawry's Green Goddess Salad Dressing Mix	1 pkg.	68.6	0.05	0.11	HC
Lawry's Hawaiian Salad Dressing	2 tbsp.	154.3	7.4	36.5	HC
Lawry's Italian Salad Dressing	2 tbsp.	158.8	0.23	6	HC
Lawry's Italian Salad Dressing Mix	1 pkg.	44	0.06	0.05	LC
Lawry's Lasagna Sauce	1 pkg.	86	4.93	6.51	HC
Lawry's Lemon Garlic Salad Dressing Mix	1 pkg.	63.8	5.1	6.2	HC
Lawry's Lemon Pepper Marinade	1 oz.	59.13	3.5	3.91	HC
Lawry's Meatloaf Sauce	1 pkg.	333.3	1.29	6.69	HC
Lawry's Mushroom Gravy Mix	1½ oz.	167	16.2	4.23	HC

Food	Quantity	Calories	Protein Grams	Carbo-hydrates Grams	P/C/C Computer
Lawry's Old-Fashioned French Salad Dressing Mix	1 tbsp.	72	0.07	0.04	HC
Lawry's Onion Salt	1 oz.	35	2.61	2.58	LC
Lawry's Packaged Season Mix	1 pkg.	120	1.05	6.17	HC
Lawry's Pinch of Herbs	1 oz.	96.77	3.05	1.28	HC
Lawry's Pork Chop'n Bag Sauce	1 pkg.	128.41	5.4	6.64	HC
Lawry's Pot Roast'n Bag Sauce	1 pkg.	121.51	10.07	5.89	HC
Lawry's Red Wine Vinegar and Oil Salad Dressing	2 tbsp.	108.1	–	27.9	HC
Lawry's San Francisco French Dressing	2 tbsp.	106.8	1.7	5.4	HC
Lawry's Seasoned Pepper	1 oz.	97.2	9.62	6.47	HC
Lawry's Seasoned Salt	1 oz.	7.1	2.33	2.74	LC
Lawry's Sherry French Dressing	2 tbsp.	110.1	1.2	10.6	HC
Lawry's Sherry Wine Sauce	1 pkg.	94.3	4.63	7.54	LP
Lawry's Sloppy Joe Packaged Season Mix	1 pkg.	138.6	8.21	6.52	HC
Lawry's Spaghetti w/Mushrooms Sauce	1 pkg.	146.6	5.10	5.26	HC
Lawry's Spanish Rice Season Mix	1 pkg.	125	1.68	4.59	HC
Lawry's Stroganoff Sauce	1 pkg.	117.7	1.29	5.1	HC
Lawry's Tartar Sauce	1 pkg.	63.9	1.63	5.43	LC
Lawry's Teriyaki BBQ Marinade Sauce	1 oz.	38.75	3.29	2.67	LC
Lawry's Thousand Island Salad Dressing	2 tbsp.	138	1.7	1.4	HC
Lawry's Thousand Island Salad Dressing Mix	1 pkg.	78	6.6	1.3	LC
Lawry's Tostada Sauce	1 pkg.	544.32	1.42	5.81	HC
Lawry's White Wine Sauce	1 pkg.	112.75	1.52	5.31	HC
Marzetti Country French Salad Dressing	1 tbsp.	71	0.26	3.55	LC
Marzetti Creamy Garlic Salad Dressing	1 tbsp.	73	0.06	1.68	LC
Marzetti Creamy Italian Salad Dressing	1 tbsp.	70	0.06	1.82	LC
Marzetti Creamy Russian Salad Dressing	1 tbsp.	88	0.08	2.06	LC
Marzetti Frenchette Gourmet Salad Dressing	1 tbsp.	21	0.11	2.2	LC
Marzetti Frenchette Green Goddess Salad Dressing	1 tbsp.	21	0.39	1.43	LC
Marzetti Frenchette Italianette Salad Dressing	1 tbsp.	6	0.05	1.46	LC

Food	Quantity	Calories	Protein Grams	Carbo-hydrates Grams	P/C/C Com-puter
Marzetti Frenchette Mayonette Gold Salad Dressing	1 tbsp.	33	0.15	1.93	LC
Marzetti Frenchette Salad Dressing	1 tbsp.	9	0.05	2.6	LC
Marzetti Frenchette Thousand Island Salad Dressing	1 tbsp.	21	0.16	3	LC
Marzetti German Style Salad Dressing	1 tbsp.	51	0.06	2.4	LC
Marzetti Horseradish	1 tbsp.	55	0.09	2.33	LC
Marzetti Low-Cal Slaw Salad Dressing	1 tbsp.	30	0.01	2.85	LC
Marzetti Low-Cal Thousand Island Salad Dressing	1 tbsp.	21	0.16	3	LC
Marzetti Low-Calorie French Dressing	1 tbsp.	9	0.05	2.6	LC
Marzetti Low-Calorie Italian Salad Dressing	1 tbsp.	6	0.05	1.46	LC
Marzetti Mint Topping Sauce	1 tbsp.	36	–	9.4	LC
Marzetti Potato Salad Dressing	1 tbsp.	62	2.88	62	HC
Marzetti Produce Roquefort Salad Dressing	1 tbsp.	68	0.61	0.72	HC
Marzetti Produce Slaw Salad Dressing	1 tbsp.	73	0.16	3.24	HC
Marzetti Produce Thousand Island Salad Dressing	1 tbsp.	69	0.12	2.25	HC
Marzetti Ranch-Style Salad Dressing	1 tbsp.	66	0.02	5.3	HC
Marzetti Slaw Salad Dressing	1 tbsp.	73	0.16	3.24	HC
Marzetti Steak Sauce	1 tbsp.	14	–	3.54	LC
Marzetti Sunny Italian Salad Dressing	1 tbsp.	79	0.03	0.35	HC
Marzetti Sweet and Saucy Salad Dressing	1 tbsp.	65	–	3.15	HC
Marzetti Tartar Sauce	1 tbsp.	69	0.08	0.82	HC
Marzetti Thousand Island Salad Dressing	1 tbsp.	68	0.11	2.28	HC
Mazola Corn Oil	1 tbsp.	120	0	0	HC
Morton Lite Salt	1 oz.	.057	0	.011	LC
Morton Plain Table Salt	1 oz.	.057	0	.011	LC
Morton Salt Substitute	1 oz.	2.26	0	0.57	LC
Morton Seasoned Salt Substitute	1 oz.	16.98	0.42	2.41	HC
Morton Table Salt, Iodized	1 oz.	.057	0	.011	LC
Mustard, prepared, brown	3½ oz.	91	5.9	5.3	HC
Mustard, prepared, yellow	3½ oz.	75	4.7	6.4	HC
Nu Made Corn Oil	1 tbsp.	125	0	0	HC

Condiments, Sauces

Food	Quantity	Calories	Protein Grams	Carbo-hydrates Grams	P/C/C Computer
Nu Made Mayonnaise	1 tbsp.	100	0	0	HC
Nu Made Safflower Oil	1 tbsp.	125	0	0	HC
Nu Made Salad Dressing	1 tbsp.	75	0	2	HC
Nu Made Vegetable Oil	1 tbsp.	125	0	0	HC
Open Pit Hickory Smoke Barbeque Sauce	1 tbsp.	27	0.02	6.5	LC
Open Pit Hot'n Spicy Barbeque Sauce	1 tbsp.	27	0.03	6.3	LC
Open Pit Original Barbeque Sauce	1 tbsp.	26	0.02	6.3	LC
Open Pit Original Barbeque Sauce w/Minced Onions	1 tbsp.	27	0.02	6.5	LC
Pantry Pride All-Vegetable Shortening	1 tbsp.	130	0	0	HC
Pantry Pride Corn Oil	1 tbsp.	125	0	0	HC
Pantry Pride Salad Oil	1 tbsp.	225	0	0	HC
Piedmont Mayonnaise	1 tbsp.	100	0	0	HC
Pillsbury Brown Gravy Mix	¼ cup	10	0	3	LC
Pillsbury Chicken Gravy Mix	¼ cup	30	1	4	LC
Pillsbury Home-Style Gravy Mix	¼ cup	10	0	3	LC
Prince Marinara Spaghetti Sauce	½ cup	70	2	9	HC
Prince Meat Spaghetti Sauce	½ cup	75	3	9	HC
Prince Meatless Spaghetti Sauce	½ cup	70	2	9	HC
Prince Spaghetti w/Mushrooms Sauce	½ cup	75	3	9	HC
Roast'n Boast Cooking Bag Beef and Sauce Packaged Mix Seasoning	1 env.	129	2.7	26	HC
Roast'n Boast Cooking Bag Chicken and Sauce Package Mix Seasoning	1 env.	119	5.5	19.3	HC
Roast'n Boast Cooking Bag Pork and Sauce Packaged Mix Seasoning	1 env.	145	6.5	27.3	HC
Roast'n Boast Cooking Bag Stew and Sauce Packaged Mix Seasoning	1 env.	126	4.6	22.6	HC
Saffola Mayonnaise	2 tbsp.	202.3	3	0.06	HC
Saffola Salad Dressing	1 oz.	98.6	0.3	4.1	HC
Saffola Salad Oil	1 oz.	252	0	0	HC
Seven Seas Creamy French Salad Dressing	1 tbsp.	60	0	2	LC
Seven Seas Creamy Italian Salad Dressing	1 tbsp.	80	0	1	LC

Food	Quantity	Calories	Protein Grams	Carbo-hydrates Grams	P/C/C Com-puter
Seven Seas Creamy Russian Salad Dressing	1 tbsp.	80	0	1	LC
Seven Seas Green Goddess Salad Dressing	1 tbsp.	70	0	0	LC
Seven Seas Low-Calorie French Dressing	1 tbsp.	30	0	2	LC
Seven Seas Low-Calorie Italian Dressing	1 tbsp.	35	0	1	LC
Seven Seas Mild Viva Italian Salad Dressing	1 tbsp.	70	0	1	LC
Seven Seas Tartar Sauce Salad Dressing	1 tbsp.	80	0	1	HC
Seven Seas Thousand Island Salad Dressing	1 tbsp.	60	0	2	LC
Seven Seas Viva Italian Salad Dressing	1 tbsp.	70	0	1	LC
Seven Seas Viva Red Wine Vinegar and Oil	1 tbsp.	70	0	1	LC
Soy sauce	3½ oz.	68	5.6	9.5	LC
Stokely-Van Camp Catsup	1 tbsp.	18.5	0.35	4.44	LC
Stokely-Van Camp Chili Sauce	1 tbsp.	14.88	0.35	3.56	LC
Stokely-Van Camp Pimientos	1 tbsp.	3.88	0.13	0.79	LC
Stokely-Van Camp Tomato Paste	1 tbsp.	13.25	0.55	3.01	LC
Tomato catsup, bottled	3½ oz.	106	2.0	25.4	HC
Tomato chili sauce, bottled	3½ oz.	104	2.5	24.8	HC
Tillie Lewis Tasti-Diet Catsup Topping	1 tbsp.	8	–	2	LC
Tillie Lewis Tasti-Diet Cheese Salad Dressing	1 tbsp.	12	0	1	LC
Tillie Lewis Tasti-Diet Chef's Salad Dressing	1 tbsp.	2	0	0	LC
Tillie Lewis Tasti-Diet French Salad Dressing	1 tbsp.	6	0	2	LC
Tillie Lewis Tasti-Diet Italian Salad Dressing	1 tbsp.	2	0	0	LC
Tillie Lewis Tasti-Diet May-Lo-Naise Dressing	1 tbsp.	16	0	1	LC
Tillie Lewis Tasti-Diet Remoulade Salad Dressing	1 tbsp.	10	0	1	LC
Tillie Lewis Tasti-Diet Whipped Salad Dressing	1 tbsp.	16	0	1	LC
Town House Meat Spaghetti Sauce	½ cup	80	2	12	HC
Town House Spaghetti w/Mushrooms Sauce	½ cup	90	2	14	HC
Town House Spaghetti Sauce	½ cup	80	2	12	HC

Food	Quantity	Calories	Protein Grams	Carbo-hydrates Grams	P/C/C Computer
Town House Tomato Paste	½ cup	112.5	4.5	26.25	HC
Town House Tomato Sauce	½ cup	45	2	9.5	LC
White sauce:					
Thin	3½ oz.	121	3.9	7.2	HC
Medium	3½ oz.	162	3.9	8.8	HC
Thick	3½ oz.	198	4.0	11.0	HC

Desserts

Food	Quantity	Calories	Protein Grams	Carbo-hydrates Grams	P/C/C Computer
Angel food cake, home recipe	3½ oz.	269	7.1	60.2	HC
Apple brown betty	3½ oz.	151	1.6	29.7	HC
Baker's Angel Flake Coconut	¼ cup	89	0.6	7.4	HC
Baker's Coconut Crunchies	¼ cup	176	1.7	10.3	HC
Baker's Cookie Coconut	¼ cup	140	1	11.6	HC
Baker's German Sweet Chocolate	4½" square	141	1.1	16.9	HC
Baker's Premium Shred Coconut	¼ cup	105	0.7	9.1	HC
Baker's Semi-Sweet Chocolate	1 square	132	1.6	16.2	HC
Baker's Semi-Sweet Chocolate Bits	¼ cup	191	1.7	28.5	HC
Baker's Southern-Style Coconut	¼ cup	85	0.7	7.4	HC
Baker's Unsweetened Chocolate	1 square	136	3.1	7.7	HC
Banquet Frozen Apple Pie	¼ pie	351	14	198	HC
Banquet Frozen Banana Cream Pie	1/5 pie	185	7	100	HC
Banquet Frozen Blackberry Pie	¼ pie	376	16	222	HC
Banquet Frozen Blueberry Pie	¼ pie	366	13	223	HC
Banquet Frozen Boysenberry Pie	¼ pie	374	16	223	HC
Banquet Frozen Butterscotch Cream Pie	1/5 pie	187	8	108	HC
Banquet Frozen Cherry Pie	¼ pie	352	16	201	HC
Banquet Frozen Chocolate Cream Pie	1/5 pie	202	10	114	HC
Banquet Frozen Coconut Cream Pie	1/5 pie	209	8	97	HC
Banquet Frozen Coconut Custard Pie	¼ pie	294	27	159	HC
Banquet Frozen Custard Pie	¼ pie	274	27	165	HC
Banquet Frozen Key Lime Pie	¼ pie	204	7	110	HC
Banquet Frozen Lemon Cream Pie	1/5 pie	179	6	102	HC
Banquet Frozen Mincemeat Pie	¼ pie	401	19	251	HC
Banquet Frozen Neopolitan Cream Pie	1/5 pie	188	8	109	HC
Banquet Frozen Peach Pie	¼ pie	320	14	182	HC
Banquet Frozen Pumpkin Pie	¼ pie	306	24	186	HC
Banquet Frozen Strawberry Cream Pie	1/5 pie	187	7	110	HC
Betty Crocker Butter Brickie Frosting Mix	1/12 pkg.	150	0	30	HC
Betty Crocker Butter Brickie Layer Cake Mix	1/12 cake	200	3	34	HC
Betty Crocker Butter Pecan Canned Frosting	1/12 can	160	0	26	HC
Betty Crocker Butter Pecan Frosting Mix	1/12 pkg.	150	0	30	HC
Betty Crocker Butter Pecan Layer Cake Mix	1/12 cake	200	3	34	HC

Food	Quantity	Calories	Protein Grams	Carbo-hydrates Grams	P/C/C Computer
Betty Crocker Caramel Apple Frosting Mix	1/12 pkg.	150	0	30	HC
Betty Crocker Cherry Canned Frosting	1/12 can	160	0	26	HC
Betty Crocker Cherry Chip Layer Cake Mix	1/12 cake	200	3	36	HC
Betty Crocker Cherry Fluff Frosting Mix	1/12 pkg.	60	0	16	HC
Betty Crocker Cherry Fudge Frosting Mix	1/12 pkg.	150	0	30	HC
Betty Crocker Cherry Fudge Layer Cake Mix	1/12 cake	200	3	34	HC
Betty Crocker Chiquita Banana Frosting Mix	1/12 pkg.	150	0	30	HC
Betty Crocker Chiquita Banana Layer Cake Mix	1/12 cake	200	4	34	HC
Betty Crocker Chocolate Angel Food Cake Mix	1/12 cake	140	3	32	HC
Betty Crocker Chocolate Canned Frosting	1/12 can	170	0	25	HC
Betty Crocker Chocolate Fluff Frosting Mix	1/12 pkg.	80	0	13	HC
Betty Crocker Chocolate Fudge Cake Mix	1/12 pkg.	160	0	32	HC
Betty Crocker Chocolate Fudge Supreme Layer Cake Mix	1/12 cake	200	3	34	HC
Betty Crocker Chocolate Malt Frosting Mix	1/12 pkg.	150	0	30	HC
Betty Crocker Chocolate Malt Layer Cake Mix	1/12 cake	200	3	34	HC
Betty Crocker Chocolate Nut Canned Frosting	1/12 can	160	0	24	HC
Betty Crocker Chocolate Walnut Frosting Mix	1/12 pkg.	150	0	30	HC
Betty Crocker Chocolate Whipped Frosting Mix	1/12 pkg.	100	0	18	HC
Betty Crocker Coconut Pecan Frosting Mix	1/12 pkg.	110	0	18	HC
Betty Crocker Confetti Angel Food Cake Mix	1/12 cake	150	3	34	HC
Betty Crocker Creamy Cherry Frosting Mix	1/12 pkg.	150	0	30	HC
Betty Crocker Creamy Spice Frosting Mix	1/12 pkg.	150	0	30	HC
Betty Crocker Creamy White Frosting Mix	1/12 pkg.	150	0	33	HC

Desserts

Food	Quantity	Calories	Protein Grams	Carbo-hydrates Grams	P/C/C Com-puter
Betty Crocker Dark Chocolate Fudge Frosting Mix	1/12 pkg.	150	0	30	HC
Betty Crocker Dark Dutch Canned Frosting	1/12 can	160	0	24	HC
Betty Crocker Devil's Food Butter Recipe	1/12 cake	280	4	35	HC
Betty Crocker Devil's Food Layer Cake Mix	1/12 cake	200	3	34	HC
Betty Crocker Dole Pineapple Frosting Mix	1/12 pkg.	150	0	30	HC
Betty Crocker Dole Pineapple Layer Cake Mix	1/12 cake	200	3	34	HC
Betty Crocker French Vanilla Layer Cake Mix	1/12 cake	200	3	34	HC
Betty Crocker Fudge Nugget Frosting Mix	1/12 pkg.	150	0	30	HC
Betty Crocker German Chocolate Layer Cake Mix	1/12 cake	200	3	34	HC
Betty Crocker Golden Caramel Frosting Mix	1/12 pkg.	150	0	30	HC
Betty Crocker Golden Pound Cake Mix	3½ oz.	190	3	28	HC
Betty Crocker Lemon Chiffon Cake Mix	1/12 cake	190	4	35	HC
Betty Crocker Lemon Custard Angel Food Cake Mix	1/12 cake	140	3	32	HC
Betty Crocker Lemon Fluff Frosting Mix	1/12 pkg.	60	0	16	LC
Betty Crocker Lemon Whipped Frosting Mix	1/12 pkg.	100	0	18	HC
Betty Crocker Marble Layer Cake Mix	1/12 cake	200	3	34	HC
Betty Crocker Milk Chocolate Canned Frosting	1/12 can	160	0	25	HC
Betty Crocker Milk Chocolate Frosting Mix	1/12 pkg.	150	0	30	HC
Betty Crocker Milk Chocolate Layer Cake Mix	1/12 cake	200	3	34	HC
Betty Crocker One-Step Angel Food Cake Mix	1/12 cake	140	3	32	HC
Betty Crocker Orange Chiffon Cake Mix	1/12 cake	190	4	35	HC
Betty Crocker Sour Cream Chocolate Canned Frosting	1/12 can	170	0	25	HC
Betty Crocker Sour Cream Chocolate Frosting Mix	1/12 pkg.	150	0	30	HC

Desserts

Food	Quantity	Calories	Protein Grams	Carbo- hydrates Grams	P/C/C Com- puter
Betty Crocker Sour Cream Chocolate Layer Cake Mix	1/12 cake	200	3	34	HC
Betty Crocker Sour Cream White Canned Frosting	1/12 can	160	0	26	HC
Betty Crocker Sour Cream White Frosting Mix	1/12 pkg.	150	0	30	HC
Betty Crocker Sour Cream White Layer Cake Mix	1/12 cake	200	3	34	HC
Betty Crocker Sour Cream Yellow Layer Cake Mix	1/12 cake	200	3	34	HC
Betty Crocker Spice 'n Apples w/Raisins Layer Cake Mix	1/12 cake	210	3	37	HC
Betty Crocker Spice Layer Cake Mix	1/12 cake	200	3	34	HC
Betty Crocker Strawberry Angel Food Cake Mix	1/12 cake	150	3	34	HC
Betty Crocker Strawberry Cream Layer Cake Mix	1/12 cake	190	3	34	HC
Betty Crocker Strawberry Whipped Frosting Mix	1/12 pkg.	100	0	18	HC
Betty Crocker Sunkist Lemon Canned Frosting	1/12 can	160	0	26	HC
Betty Crocker Sunkist Lemon Frosting Mix	1/12 pkg.	150	0	30	HC
Betty Crocker Sunkist Lemon Layer Cake Mix	1/12 cake	200	3	34	HC
Betty Crocker Sunkist Orange Canned Frosting	1/12 can	160	0	26	HC
Betty Crocker Sunkist Orange Frosting Mix	1/12 pkg.	150	0	30	HC
Betty Crocker Sunkist Orange Layer Cake Mix	1/12 cake	200	3	34	HC
Betty Crocker Toasted Coconut Frosting Mix	1/12 pkg.	150	0	30	HC
Betty Crocker Traditional Angel Food Cake Mix	1/12 cake	130	3	30	HC
Betty Crocker Vanilla Canned Frosting	1/12 can	160	0	26	HC
Betty Crocker Vanilla Whipped Frosting Mix	1/12 pkg.	100	0	18	HC
Betty Crocker White Fluff Frosting Mix	1/12 pkg.	60	0	16	HC
Betty Crocker White Layer Cake Mix	1/12 cake	200	3	34	HC
Betty Crocker Yellow Butter Recipe	1/12 cake	280	4	35	HC

Food	Quantity	Calories	Protein Grams	Carbo-hydrates Grams	P/C/C Computer
Betty Crocker Yellow Layer Cake Mix	1/12 cake	200	3	34	HC
Birds Eye Cool 'n Creamy Pudding, All Flavors	½ cup	172	2.6	27.7	HC
Birds Eye Cool Whip Dessert Topping	1 tbsp.	16	–	1.1	LC
Boston cream pie	3½ oz.	302	5.0	49.9	HC
Bread pudding with raisins	3½ oz.	187	5.6	28.4	HC
Caramel cake icing	3½ oz.	360	1.3	76.5	HC
Carnation Chocolate Instant Breakfast Pudding	½ pkg. + 4 oz. milk	290	16	32	HC
Carnation Vanilla Instant Breakfast Pudding	½ pkg. + 4 oz. milk	290	16	33	HC
Carvel Chocolate Ice Cream	3 oz.	147	3.19	16.79	HC
Carvel Sherbert	3 oz.	105	1.06	23.37	HC
Carvel Vanilla Ice Cream	3 oz.	148	3.32	15.91	HC
Chocolate cake icing	3½ oz.	376	3.2	67.4	HC
Chocolate, devil's food	3½ oz.	369	4.5	55.8	HC
Chocolate fudge cake icing	3½ oz.	378	2.2	67.0	HC
Chocolate malt cake	3½ oz.	346	3.4	66.6	HC
Coconut cake icing	3½ oz.	364	1.9	74.9	HC
Coffee cake	3½ oz.	322	6.3	52.4	HC
Corn pudding	3½ oz.	104	4.0	13.0	HC
Cottage pudding	3½ oz.	318	5.3	56.7	HC
Cream puffs with custard filling	3½ oz.	233	6.5	20.5	HC
Creamy fudge cake icing	3½ oz.	383	2.6	65.9	HC
Cupcake	3½ oz.	368	4.2	59.4	HC
Del Monte Dessert Cup Pudding 'n Apricot	5 oz.	173	1.1	34.9	HC
Del Monte Dessert Cup Pudding 'n Peach	5 oz.	173	1.0	35.5	HC
Del Monte Dessert Cup Pudding 'n Pineapple	5 oz.	173	1.0	34.6	HC
Del Monte Gel Cup Lemon-Lime w/Pineapple	5 oz.	115	0.3	28.5	HC
Del Monte Gel Cup Orange w/Peaches Gelatin	5 oz.	107	0.4	26.5	HC
Del Monte Gel Cup Strawberry w/Peaches Gelatin	5 oz.	111	0.4	27.4	HC
Del Monte Pudding Cup, Butterscotch	5 oz.	191	3.3	32.6	HC
Del Monte Pudding Cup, Chocolate	5 oz.	202	4.2	33.6	HC
Del Monte Pudding Cup, Chocolate Fudge	5 oz.	198	4.2	32.4	HC

Desserts

Food	Quantity	Calories	Protein Grams	Carbo-hydrates Grams	P/C/C Computer
Del Monte Pudding Cup, Vanilla	5 oz.	190	3.1	32.8	HC
Devil's food (frozen) cake	3½ oz.	371	3.5	43.8	HC
Dream Whip Whipped Dessert Topping	1 tbsp.	14	0.2	1.2	LC
Dromedary Pound Cake Mix	½ cake	385	5.2	50.1	HC
D-Zerta, All Flavors	½ cup	8	8	–	LC
D-Zerta Butterscotch w/Nonfat Milk Pudding	½ cup	72	4.8	12.7	LC
D-Zerta Butterscotch w/Whole Milk Pudding	½ cup	107	4.8	12.2	HC
D-Zerta Chocolate w/Nonfat Milk Pudding	½ cup	67	5	11.6	LC
D-Zerta Chocolate w/Whole Milk Pudding	½ cup	102	5	11	HC
D-Zerta Vanilla w/Nonfat Milk Pudding	½ cup	72	4.8	12.7	LC
D-Zerta Vanilla w/Whole Milk Pudding	½ cup	107	4.8	12.2	HC
D-Zerta Whipped Dessert Topping	1 tbsp.	7	0.1	0.3	LC
Flako Cupcake Pastry Mix	1	140	1.9	21.9	HC
Flako Pie Crust Mix	1/6 pie	232	3.2	23.4	HC
Fondant	3½ oz.	364	2.0	89.6	HC
Fruitcake	3½ oz.	379	4.8	59.7	HC
Gingerbread	3½ oz.	317	3.8	52.0	HC
Honey spice cake	3½ oz.	352	4.1	60.9	HC
Hostess Apple Fruit Pie	1	352.35	4.01	60.36	HC
Hostess Big Wheels	1	166.26	1.74	21.57	HC
Hostess Brownies	1	100.14	1.46	15.50	HC
Hostess Cake Donuts	1	138.56	1.63	18.22	HC
Hostess Cherry Fruit Pie	1	336.58	4.10	56.03	HC
Hostess Devil's Food Cupcake	1	160.13	1.70	29.57	HC
Hostess Dark Ding Dongs	1	166.65	1.34	21.63	HC
Hostess Ho Hos	1	102.84	0.81	14.65	HC
Hostess Lemon Fruit Pie	1	357.25	4.39	53.92	HC
Hostess Milk Ding Dongs	1	166.26	1.74	21.57	HC
Hostess Orange Cupcake	1	149.78	1.54	26.72	HC
Hostess Snowballs	1	135.04	1.33	25.75	HC
Hostess Suzy Q's	1	231.12	2.08	36.54	HC
Hostess Twinkies	1	143.85	1.79	24.96	HC
Hunt's Snack Pack Butterscotch	1 can	238	2.18	30.3	HC
Hunt's Snack Pack Chocolate	1 can	239	2.43	30.3	HC
Hunt's Snack Pack Cherry Gelatin	5 oz.	89.4	–	22.5	HC
Hunt's Snack Pack Chocolate Fudge	1 can	229	2.21	28.7	HC
Hunt's Snack Pack Lemon Pudding	1 can	175	0.03	35.3	HC

Food	Quantity	Calories	Protein Grams	Carbo-hydrates Grams	P/C/C Computer
Hunt's Snack Pack Orange Gelatin	5 oz.	105	–	26.6	HC
Hunt's Snack Pack Raspberry Gelatin	5 oz.	105	–	26.6	HC
Hunt's Snack Pack Strawberry Gelatin	5 oz.	106	–	26.82	HC
Hunt's Snack Pack Vanilla Pudding	1 can	238	2.36	30.2	HC
Jams and preserves	3½ oz.	272	.6	70.0	HC
Jellies	3½ oz.	273	.1	70.6	HC
Jell-O Banana w/Nonfat Milk Pudding	½ cup	136	4.4	29.5	HC
Jell-O Banana w/Whole Milk Pudding	½ cup	173	4.3	29.3	HC
Jell-O Butterscotch w/Nonfat Milk Pudding	½ cup	136	4.4	29.5	HC
Jell-O Butterscotch w/Whole Milk Pudding	½ cup	173	4.3	29.3	HC
Jell-O Cheesecake Dessert	1/8 cake	255	5.9	31.4	HC
Jell-O Chocolate Fudge w/Nonfat Milk Pudding	½ cup	139	5.2	29.7	HC
Jell-O Chocolate Fudge w/Whole Milk Pudding	½ cup	174	5.0	29.5	HC
Jell-O Chocolate w/Nonfat Milk Pudding	½ cup	139	5.2	29.7	HC
Jell-O Chocolate w/Whole Milk Pudding	½ cup	174	5	29.5	HC
Jell-O Coconut Cream w/Nonfat Milk Pudding	½ cup	135	4.6	26	HC
Jell-O Coconut Cream w/Whole Milk Pudding	½ cup	175	4.5	25.8	HC
Jell-O French Vanilla w/Nonfat Milk Pudding	½ cup	136	4.4	29.5	HC
Jell-O French Vanilla w/Whole Milk Pudding	½ cup	173	4.3	29.3	HC
Jell-O Golden Egg w/Nonfat Milk Pudding	½ cup	128	5.9	23.1	HC
Jell-O Golden Egg w/Whole Milk Pudding	½ cup	165	5.8	22.9	HC
Jell-O Instant Banana Pudding	½ cup	178	4.3	30.5	HC
Jell-O Instant Butterscotch Pudding	½ cup	178	4.3	30.5	HC
Jell-O Instant Chocolate Pudding	½ cup	190	5.2	33.4	HC
Jell-O Instant Chocolate Fudge Pudding	½ cup	190	5.2	33.4	HC

Food	Quantity	Calories	Protein Grams	Carbo-hydrates Grams	P/C/C Computer
Jell-O Instant Coconut Cream Pudding	½ cup	188	4.5	29	HC
Jell-O Instant French Vanilla Pudding	½ cup	178	4.3	30.5	HC
Jell-O Instant Lemon Pudding	½ cup	178	4.3	30.5	HC
Jell-O Instant Pineapple Pudding	½ cup	178	4.3	30.5	HC
Jell-O Instant Vanilla Pudding	½ cup	178	4.3	30.5	HC
Jell-O Lemon Pudding	½ cup	176	2	38.8	HC
Jell-O Milk Chocolate w/Nonfat Milk Pudding	½ cup	139	5.2	29.7	HC
Jell-O Milk Chocolate w/Whole Milk Pudding	½ cup	174	5	29.5	HC
Jell-O 1-2-3, All Flavors	2/3 cup	135	1.5	25.2	HC
Jell-O Regular, All Flavors	½ cup	81	1.7	18.2	LC
Jell-O Tapioca, All Flavors	½ cup	166	4.6	27.6	HC
Jell-O Vanilla w/Nonfat Milk Pudding	½ cup	136	4.4	29.5	HC
Jell-O Vanilla w/Whole Milk Pudding	½ cup	173	4.3	29.3	HC
Jell-O Wild, All Flavors	½ cup	81	1.7	17.8	LC
Jujube (Chinese date) dried	3½ oz.	287	3.7	73.6	HC
Kool-Pop Bars Frozen Dessert	1 bar	32	–	8.1	LC
Marble cake	3½ oz.	331	4.4	62.0	HC
Minute Tapioca	½ cup	150	5.9	20.1	HC
Morton Frozen Apple Pie	1/6 pie	239.56	2.17	33.64	HC
Morton Frozen Banana Cream Pie	¼ pie	256.06	2.19	34.91	HC
Morton Frozen Blueberry Pie	1/6 pie	240.37	2.27	33.53	HC
Morton Frozen Cherry Pie	1/6 pie	249.78	2.45	35.71	HC
Morton Frozen Coconut Cream Pie	¼ pie	278.45	2.22	37.08	HC
Morton Frozen Coconut Custard Pie	1/6 pie	203.25	2.87	27.70	HC
Morton Frozen German Chocolate Cake	1/12 cake	230	2.9	28	HC
Morton Frozen Lemon Cream Pie	¼ pie	261.86	2.07	36.40	HC
Morton Frozen Mincemeat Pie	1/6 pie	246.92	2.21	35.83	HC
Morton Frozen Neopolitan Cream Pie	¼ pie	267.20	2.28	35.42	HC
Morton Frozen Peach Pie	1/6 pie	245.28	2.23	34.94	HC
Morton Frozen Pecan Pie	1/6 pie	343.28	3.72	49.42	HC
Morton Frozen Pineapple Pie	1/8 pie	399.18	2.99	60.56	HC
Morton Frozen Pound Cake	3½ oz.	117.3	1.57	14.8	HC
Morton Frozen Pumpkin Pie	1/6 pie	166.71	2.85	26.15	HC
Morton Frozen Rhubarb-Strawberry Pie	1/8 pie	433.41	3.21	68.78	HC
Morton Frozen Strawberry Pie	1/6 pie	252.67	2.29	36.75	HC

Food	Quantity	Calories	Protein Grams	Carbo-hydrates Grams	P/C/C Computer
Morton Frozen Strawberry Cream Pie	1/6 pie	247.26	2.28	32.40	HC
Pantry Pride Whipped Dessert Topping	¼ oz.	24	0.09	1.7	LC
Pepperidge Farm Frozen Chocolate Fudge Cake	1/6 cake	315	3.6	43.4	HC
Pepperidge Farm Frozen Coconut Cake	1/6 cake	323	2.7	46	HC
Pepperidge Farm Frozen Devil's Food Cake	1/6 cake	325	3.1	47	HC
Pepperidge Farm Frozen Golden Cake	1/6 cake	320	3	43.6	HC
Pepperidge Farm Patty Shell Pie Crust	1 shell	232	2.5	14.5	HC
Pies, baked, piecrust with un-enriched flour:					
Apple	3½ oz.	256	2.2	38.1	HC
Banana custard	3½ oz.	221	4.5	30.7	HC
Blackberry	3½ oz.	243	2.6	34.4	HC
Blueberry	3½ oz.	242	2.4	34.9	HC
Butterscotch	3½ oz.	267	4.4	38.3	HC
Cherry	3½ oz.	261	2.6	38.4	HC
Chocolate chiffon	3½ oz.	328	6.8	43.7	HC
Chocolate meringue	3½ oz.	252	4.8	33.5	HC
Coconut custard	3½ oz.	235	6.0	24.9	HC
Custard	3½ oz.	218	6.1	23.4	HC
Lemon chiffon	3½ oz.	313	7.0	43.8	HC
Lemon meringue	3½ oz.	255	3.7	37.7	HC
Mince	3½ oz.	271	2.5	41.2	HC
Peach	3½ oz.	255	2.5	38.2	HC
Pecan	3½ oz.	418	5.1	51.3	HC
Pineapple	3½ oz.	253	2.2	38.1	HC
Pineapple chiffon	3½ oz.	288	6.6	39.1	HC
Pineapple custard	3½ oz.	220	4.0	32.1	HC
Pumpkin	3½ oz.	211	4.0	24.5	HC
Raisin	3½ oz.	270	2.6	43.0	HC
Rhubarb	3½ oz.	253	2.5	38.2	HC
Strawberry	3½ oz.	198	1.9	30.9	HC
Sweet potato	3½ oz.	213	4.5	23.7	HC
Pie mix and pie baked from mix	3½ oz.	470	3.3	70.6	HC
Pie mix with egg yolk-milk, baked	3½ oz.	203	4.3	29.1	HC
Piecrust or plain pastry	3½ oz.	500	6.1	43.8	HC
Pillsbury Apple Fruit 'n Crunch Bar Pastry Mix	2" square	140	2	18	HC
Pillsbury Apple Turnover Pastry Mix	1	150	1	23	HC

Desserts

Food	Quantity	Calories	Protein Grams	Carbo-hydrates Grams	P/C/C Com-puter
Pillsbury Applesauce Spice Layer Cake Mix	1/12 cake	200	3	36	HC
Pillsbury Banana Layer Cake Mix	1/12 cake	190	3	34	HC
Pillsbury Blueberry Fruit 'n Crunch Bar Pastry Mix	2" square	140	2	18	HC
Pillsbury Blueberry Turnovers Pastry Mix	1	150	2	23	HC
Pillsbury Butter Layer Cake Mix	1/12 cake	210	3	34	HC
Pillsbury Butterscotch Swirl Bundt Cake	1/12 cake	310	4	48	HC
Pillsbury Caramel Frosting Mix	1/12 pkg.	160	0	28	HC
Pillsbury Cherry Fruit 'n Crunch Bar Pastry Mix	2" square	140	2	18	HC
Pillsbury Cherry Turnovers Pastry Mix	1	150	2	23	HC
Pillsbury Chocolate Fudge Cake Mix	1/12 pkg.	160	0	28	HC
Pillsbury Chocolate Fudge Canned Frosting	1/12 can	150	0	29	HC
Pillsbury Chocolate Fudge Layer Cake Mix	1/12 cake	210	3	34	HC
Pillsbury Chocolate Macaroon Bundt Cake	1/12 cake	360	4	52	HC
Pillsbury Cinnamon Swirl Streusel Cake Mix	1/12 cake	350	4	52	HC
Pillsbury Coconut Almond Frosting Mix	1/12 pkg.	170	2	17	HC
Pillsbury Coconut Pecan Frosting Mix	1/12 pkg.	150	1	20	HC
Pillsbury Devil's Food Cupcake Pastry Mix	1	170	2	30	HC
Pillsbury Double Dutch Canned Frosting	1/12 can	150	0	29	HC
Pillsbury Double Dutch Frosting Mix	1/12 pkg.	140	0	27	HC
Pillsbury Double Dutch Layer Cake Mix	1/12 cake	210	4	33	HC
Pillsbury Fluffy White Frosting Mix	1/12 pkg.	70	0	17	HC
Pillsbury Fudge Brownie Pastry Mix	2, each 1½" sq.	120	1	19	HC
Pillsbury Fudge Cupcake Pastry Mix	1	180	2	32	HC
Pillsbury Fudge Macaroon Layer Cake Mix	1/12 cake	200	3	33	HC
Pillsbury Fudge Marble Streusel Cake Mix	1/12 cake	350	4	52	HC

Food	Quantity	Calories	Protein Grams	Carbo-hydrates Grams	P/C/C Computer
Pillsbury Fudge Nut Crown Bundt Cake	1/12 cake	300	4	43	HC
Pillsbury German Chocolate Layer Cake Mix	1/12 cake	210	3	34	HC
Pillsbury German Chocolate Whirl Streusel Cake Mix	1/12 cake	350	4	52	HC
Pillsbury Gingerbread Pastry Mix	3" square	190	2	36	HC
Pillsbury Lemon Blueberry Bundt Cake	1/12 cake	310	4	47	HC
Pillsbury Lemon Canned Frosting	1/12 can	150	0	29	HC
Pillsbury Lemon Frosting Mix	1/12 pkg.	160	0	28	HC
Pillsbury Lemon Layer Cake Mix	1/12 cake	210	3	34	HC
Pillsbury Lemon Swirl Streusel Cake Mix	1/12 cake	350	4	52	HC
Pillsbury Light Chocolate Canned Frosting	1/12 can	150	0	29	HC
Pillsbury Milk Chocolate Frosting Mix	1/12 pkg.	150	0	26	HC
Pillsbury Peach Fruit 'n Crunch Bar Pastry Mix	2" square	140	2	19	HC
Pillsbury Pecan Bar Pastry Mix	2, each 1¼" sq.	190	1	21	HC
Pillsbury Pie Crust Mix	1/6 pie	290	4	27	HC
Pillsbury Pineapple Layer Cake Mix	1/12 cake	210	3	34	HC
Pillsbury Pound Bundt Cake	1/12 cake	300	4	45	HC
Pillsbury Raspberry Angel Food Cake Mix	1/12 cake	140	3	33	HC
Pillsbury Raspberry Swirl Bundt Cake	1/12 cake	290	4	45	HC
Pillsbury Red Devil's Food Layer Cake Mix	1/12 cake	210	3	33	HC
Pillsbury Rich Chocolate Swirl Streusel Cake Mix	1/12 cake	340	4	52	HC
Pillsbury Sour Cream Chocolate Layer Cake Mix	1/12 cake	210	4	33	HC
Pillsbury Spice Swirl Streusel Cake Mix	1/12 cake	350	4	52	HC
Pillsbury Sticks Pie Crust	1/6 pie	290	4	27	HC
Pillsbury Strawberry Canned Frosting	1/12 can	150	0	29	HC
Pillsbury Strawberry Frosting Mix	1/12 pkg.	150	0	27	HC
Pillsbury Strawberry Fruit 'n Crunch Bar Pastry Mix	2" square	140	2	19	HC
Pillsbury Strawberry Layer Cake Mix	1/12 cake	190	3	34	HC

Desserts

Food	Quantity	Calories	Protein Grams	Carbohydrates Grams	P/C/C Computer
Pillsbury Vanilla Canned Frosting	1/12 can	150	0	29	HC
Pillsbury Vanilla Frosting Mix	1/12 pkg.	160	0	29	HC
Pillsbury Walnut Brownie Pastry Mix	2, each 1½" sq.	130	1	20	HC
Pillsbury Whipping Cream White Layer Cake Mix	1/12 cake	230	4	35	HC
Pillsbury White Angel Food Cake Mix	1/12 cake	140	3	33	HC
Pillsbury White Layer Cake Mix	1/12 cake	200	3	34	HC
Pillsbury Yellow Cupcake Pastry Mix	1	170	2	31	HC
Pound cake	3½ oz.	411	6.4	54.7	HC
Prune whip	3½ oz.	156	4.4	36.9	HC
Puddings prepared from home recipe:					
Chocolate	3½ oz.	148	3.1	25.7	HC
Vanilla (blancmange)	3½ oz.	111	3.5	15.9	HC
Pudding mixes and puddings made from mixes	3½ oz.	124	3.4	22.8	HC
R & R Plum Pudding	3½ oz.	266	3.6	60.1	HC
Rennin products:					
Tablet (salts, starch, renin enzyme)	3½ oz.	107	.1	24.3	HC
Chocolate dessert mix with milk	3½ oz.	102	3.4	14.1	HC
Dessert, home-prepared with tablet	3½ oz.	89	3.1	11.6	HC
Other flavors dessert mix with milk	3½ oz.	95	3.2	12.8	HC
Rice pudding with raisins	3½ oz.	146	3.6	26.7	HC
Sherbert, orange	3½ oz.	134	.9	30.8	HC
Smucker's Butterscotch Dessert Topping	2 tbsp.	140	0	33	HC
Smucker's Caramel Dessert Topping	2 tbsp.	140	1	33	HC
Smucker's Cherry Dessert Topping	2 tbsp.	130	0	32	HC
Smucker's Chocolate Dessert Topping	2 tbsp.	130	1	27	HC
Smucker's Chocolate Fudge Dessert Topping	2 tbsp.	130	1	31	HC
Smucker's Chocolate-Mint Fudge Dessert Topping	2 tbsp.	140	2	31	HC

Food	Quantity	Calories	Protein Grams	Carbo-hydrates Grams	P/C/C Computer
Smucker's Peanut Butter Caramel Dessert Topping	2 tbsp.	150	3	29	HC
Smucker's Pecan Dessert Topping	2 tbsp.	150	2	33	HC
Smucker's Pineapple Dessert Topping	2 tbsp.	130	0	32	HC
Smucker's Strawberry Dessert Topping	2 tbsp.	120	0	30	HC
Smucker's Swiss Milk Chocolate Dessert Topping	2 tbsp.	140	3	31	HC
Smucker's Walnut Dessert Topping	2 tbsp.	150	2	32	HC
Soft Swirl w/Nonfat Milk Chocolate Dessert	½ cup	163	4.2	27.6	HC
Soft Swirl w/Whole Milk Chocolate Dessert	½ cup	190	4.1	27.4	HC
Soft Swirl w/Nonfat Milk Peach Dessert	½ cup	139	3.2	26	HC
Soft Swirl w/Whole Milk Peach Dessert	½ cup	168	3.1	25.8	HC
Soft Swirl w/Nonfat Milk Strawberry Dessert	½ cup	139	3.2	26	HC
Soft Swirl w/Whole Milk Strawberry Dessert	½ cup	168	3.1	25.8	HC
Soft Swirl w/Nonfat Milk Vanilla Dessert	½ cup	139	3.2	26	HC
Soft Swirl w/Whole Milk Vanilla Dessert	½ cup	168	3.1	25.8	HC
Sponge cake	3½ oz.	297	7.6	54.1	HC
Swans Down Devil's Food Layer Cake Mix	1/12 cake	184	3.3	35.3	HC
Swans Down White Angel Food Cake Mix	1/12 cake	132	3.2	29.7	HC
Swans Down Yellow Layer Cake Mix	1/12 cake	186	3.1	36.1	HC
Tapioca, dry	3½ oz.	352	.6	86.4	HC
Tapioca desserts:					
Apple tapioca	3½ oz.	117	.2	29.4	HC
Tapioca cream pudding	3½ oz.	134	5.0	17.1	HC
Tartar sauce	3½ oz.	531	1.4	4.2	HC
Thinny-Thin Frozen Chocolate Dessert	5.9 oz.	109	4.5	20.5	LC
Thinny-Thin Frozen Coffee Dessert	5.9 oz.	109	5	20	LC
Thinny-Thin Frozen Vanilla Dessert	5.9 oz.	109	5	20	LC
Tillie Lewis Tasti-Diet Apple Jelly	1 tsp.	4	0	1	LC

Desserts

Food	Quantity	Calories	Protein Grams	Carbo-hydrates Grams	P/C/C Com-puter
Tillie Lewis Tasti-Diet Apricot-Pineapple Jelly	1 tsp.	4	0	1	LC
Tillie Lewis Tasti-Diet Chocolate Topping	1 tbsp.	8	–	2	LC
Tillie Lewis Tasti-Diet Pancake and Waffle Topping	1 tbsp.	4	–	1	LC
Tillie Lewis Tasti-Diet Peach Jelly	1 tsp.	4	0	1	LC
Tillie Lewis Tasti-Diet Strawberry Jelly	1 tsp.	4	0	1	LC
Whip 'n Chill w/Nonfat Milk Chocolate Dessert	½ cup	134	3.2	22.1	HC
Whip 'n Chill w/Whole Milk Chocolate Dessert	½ cup	144	3.2	22	HC
Whip 'n Chill w/Nonfat Milk Lemon Dessert	½ cup	124	2.8	19.4	HC
Whip 'n Chill w/Whole Milk Lemon Dessert	½ cup	135	2.8	19.3	HC
Whip 'n Chill w/Nonfat Milk Strawberry Dessert	½ cup	134	3.2	22.1	HC
Whip 'n Chill w/Whole Milk Strawberry Dessert	½ cup	135	2.8	19.3	HC
Whip 'n Chill w/Nonfat Milk Vanilla Dessert	½ cup	124	2.8	19.4	HC
Whip 'n Chill w/Whole Milk Vanilla Dessert	½ cup	135	2.8	19.3	HC
White cake	3½ oz.	371	3.7	60.7	HC
White cake icing	3½ oz.	316	1.4	80.3	HC
Wilderness Apple Fruit Filling	21 oz.	110	0	30	HC
Wilderness Apple French Fruit Filling	21 oz.	110	0	28	HC
Wilderness Apricot Fruit Filling	21 oz.	120	0	30	HC
Wilderness Blueberry Fruit Filling	21 oz.	110	0	28	HC
Wilderness Cherry Fruit Filling	21 oz.	110	0	29	HC
Wilderness Lemon Fruit Filling	22 oz.	140	0	39	HC
Wilderness Mince Fruit Filling	22 oz.	200	0	45	HC
Wilderness Peach Fruit Filling	21 oz.	110	0	25	HC
Wilderness Raisin Fruit Filling	22 oz.	120	0	31	HC
Wilderness Strawberry Fruit Filling	21 oz.	120	0	31	HC
Yellow cake	3½ oz.	362	4.0	61.3	HC

Snack Foods

Food	Quantity	Calories	Protein Grams	Carbo- hydrates Grams	P/C/C Com- puter
Aunt Jemima Cinnamon Sticks Pastry	3	145	3.2	21.3	HC
Aunt Jemima Coffee Cake Mix	1/8 cake	186	3	30.3	HC
Butterscotch candy	3½ oz.	397	Trace	94.8	HC
Carmels:					
Chocolate-flavored roll	3½ oz.	396	2.2	82.7	HC
Plain or chocolate	3½ oz.	399	4.0	76.6	HC
Plain or chocolate with nuts	3½ oz.	428	4.5	70.5	HC
Charlotte russe, with ladyfingers, whipped cream filling	3½ oz.	286	5.9	33.5	HC
Chewing gum	3½ oz.	317	–	95.2	HC
Chocolate, bitter or baking	3½ oz.	505	10.7	28.9	HC
Chocolate, bittersweet	3½ oz.	477	7.9	46.8	HC
Chocolate-coated:					
Almonds	3½ oz.	569	12.3	39.6	HC
Chocolate fudge	3½ oz.	430	3.8	73.1	HC
Chocolate fudge with nuts	3½ oz.	452	4.9	67.3	HC
Coconut center	3½ oz.	438	2.8	72.0	HC
Fondant	3½ oz.	410	1.7	81.0	HC
Fudge, peanuts, caramel	3½ oz.	459	9.4	58.7	HC
Honeycombed hard candy with peanut butter	3½ oz.	463	6.6	70.6	HC
Nougat and caramel	3½ oz.	416	4.0	72.8	HC
Peanuts	3½ oz.	561	16.4	39.1	HC
Raisins	3½ oz.	425	5.4	70.5	HC
Vanilla creams	3½ oz.	435	3.8	70.3	HC
Chocolate discs, sugar-coated	3½ oz.	466	5.2	72.7	HC
Chocolate, milk with almonds	3½ oz.	532	9.3	51.3	HC
Chocolate, milk with peanuts	3½ oz.	543	14.1	44.6	HC
Chocolate, semisweet	3½ oz.	507	4.2	57.0	HC
Chocolate, sweet	3½ oz.	528	4.4	57.9	HC
Chocolate syrup, fudge type	3½ oz.	330	5.1	54.0	HC
Chocolate syrup, thin type	3½ oz.	245	2.3	62.7	HC
Citron, candied	3½ oz.	314	.2	80.2	HC
Cookies:					
Assorted, packaged, commercial	3½ oz.	480	5.1	71.0	HC
Brownies, frozen with chocolate icing	3½ oz.	419	4.9	60.7	HC
Brownies with nuts, home recipe	3½ oz.	485	6.5	50.9	HC

Food	Quantity	Calories	Protein Grams	Carbo-hydrates Grams	P/C/C Com-puter
Cookies *(cont.):*					
Butterthin	3½ oz.	457	6.1	70.9	HC
Chocolate	3½ oz.	445	7.1	71.5	HC
Chocolate chip	3½ oz.	471	5.4	69.7	HC
Coconut bars	3½ oz.	494	6.2	63.9	HC
Fig bars	3½ oz.	358	3.9	75.4	HC
Gingersnaps	3½ oz.	420	5.5	79.8	HC
Ladyfingers	3½ oz.	360	7.8	64.5	HC
Macaroons	3½ oz.	475	5.3	66.1	HC
Marshmallow	3½ oz.	409	4.0	72.3	HC
Molasses	3½ oz.	422	6.4	76.0	HC
Oatmeal with raisins	3½ oz.	451	6.2	73.5	HC
Peanut	3½ oz.	473	10.0	67.0	HC
Raisin	3½ oz.	379	4.4	80.8	HC
Sandwich type	3½ oz.	495	4.8	69.3	HC
Shortbread	3½ oz.	498	7.2	65.1	HC
Sugar, soft, thick	3½ oz.	444	6.0	68.0	HC
Sugar wafers	3½ oz.	485	4.9	73.4	HC
Vanilla wafers	3½ oz.	462	5.4	74.4	HC
Doughnuts, cake type	3½ oz.	391	4.6	51.4	HC
Doughnuts, yeast-leavened	3½ oz.	414	6.3	37.7	HC
Eclairs with custard filling and chocolate icing	3½ oz.	239	6.2	23.2	HC
Fudge:					
Chocolate	3½ oz.	400	2.7	75.0	HC
Chocolate with nuts	3½ oz.	426	3.9	69.0	HC
Vanilla	3½ oz.	398	3.0	74.8	HC
Vanilla, with nuts	3½ oz.	424	4.2	68.8	HC
Gumdrops, hard	3½ oz.	386	0	97.2	HC
Gumdrops, starch jelly pieces	3½ oz.	347	.1	87.4	HC
Ice cream cones	3½ oz.	377	10.0	77.9	HC
Ices, water, lime	3½ oz.	78	.4	32.6	LC
Jelly beans	3½ oz.	367	Trace	93.1	HC
Lawry's Bacon Onion Dip Mix	1 pkg.	46.9	2.34	44.97	HCR
Lawry's Bleu Cheese Dip Mix	1 pkg.	47.71	1.9	50.76	HCR
Lawry's Chili con Queso Dip Mix	1 pkg.	71.77	1.93	46.58	HCR
Lawry's Green Onion Dip Mix	1 pkg.	49.02	1.05	67.35	HCR
Lawry's Guacamole Dip Mix	1 pkg.	60.2	1.21	34.66	HCR
Lawry's Taco Dip Mix	1 pkg.	43.05	1.25	62.35	HCR
Lawry's Toasted Onion Dip Mix	1 pkg.	47.41	8.5	62.31	HCR
Marmalade, citrus	3½ oz.	257	.5	70.1	HC
Marshmallows	3½ oz.	319	2.0	80.4	HC
Morton Frozen Apple Danish Coffee Cake	Complete	1129	13.11	166.6	HC
Morton Frozen Danish Pecan Twist	Complete	1369	16.69	149	HC

Food	Quantity	Calories	Protein Grams	Carbo- hydrates Grams	P/C/C Com- puter
Morton Frozen Honey Buns	1	170	3.46	24.82	HC
Morton Frozen Melt-a-Way Coffee Cake	Complete	1511	19.16	184.8	HC
Morton Frozen Powdered Donut	1	82.1	0.84	9.25	HC
Morton Frozen Sugar and Spice Donut	1	81.9	0.76	8.62	HC
Nabisco Almond Crescent Cookies	14	464	5.4	68.9	HC
Nabisco American Harvest Crackers	31	503	8	61.8	HC
Nabisco Bacon-Flavored Crackers	48	507	8.4	59.1	HC
Nabisco Bake Shop Macaroons	5	459	3.3	64	HC
Nabisco Bake Shop Oatmeal Raisin Cookies	6	438	5.5	65.6	HC
Nabisco Barnum's Animal Crackers	39	444	5.9	76.5	HC
Nabisco Biscos Sugar Wafers	28	516	3.4	70.4	HC
Nabisco Biscos Waffle Cremes	12	520	2.6	71.5	HC
Nabisco Bridge Mix	50	442	5.2	68.7	HC
Nabisco Brown Edge Wafers	17	488	4.8	71	HC
Nabisco Brown Sugar Family Favorite Cookies	21	527	4.6	63.7	HC
Nabisco Butter-Flavored Cookies	20	476	5.3	73.4	HC
Nabisco Butter Thins Crackers	30	460	7.4	71.7	HC
Nabisco Butterscotch Skimmers	17	410	0.1	94.8	HC
Nabisco Buttery Sesame Crackers	31	503	8.3	58.3	HC
Nabisco Cameo Creme Cookies	7	478	4.9	74	HC
Nabisco Cheese Flings Crackers	56	597	9.4	37.9	HC
Nabisco Cheese Nips Crackers	91	458	9.3	66.6	HC
Nabisco Cheese Tid-Bit Crackers	111	470	8.4	61.5	HC
Nabisco Chicken in a Biskit Crackers	50	517	7.3	60.9	HC
Nabisco Chippers Crackers	36	495	6.5	64.5	HC
Nabisco Chips Ahoy Cookies	9	483	4.9	71	HC
Nabisco Chipsters Crackers	200	471	4.5	67.2	HC
Nabisco Choco-Flavored Sugar Daddy	8	405	5.7	84.4	HC
Nabisco Chocolate Cherries	6	394	1.3	78.1	HC
Nabisco Chocolate Chip Cookies	15	503	4.6	67.5	HC
Nabisco Chocolate Chip Creme	7	515	4.1	64.1	HC
Nabisco Chocolate Chip Family Favorite Cookies	15	503	4.6	67.5	HC
Nabisco Chocolate Chip Snaps	22	468	5.5	75.3	HC
Nabisco Chocolate-Covered Grahams	9	511	6	65.2	HC
Nabisco Chocolate-Covered Peanuts	24	594	16.4	39.1	HC

Snack Foods

Food	Quantity	Calories	Protein Grams	Carbo-hydrates Grams	P/C/C Computer
Nabisco Chocolate-Covered Raisins	111	458	5.4	70.5	HC
Nabisco Chocolate Snaps	26	467	6.2	71.7	HC
Nabisco Chuckles Eggs	40	385	–	92.8	HC
Nabisco Chuckles Jelly Rings	9	336	–	82.8	HC
Nabisco Chuckles Ju-Jubes	27	362	–	88.3	HC
Nabisco Chuckles Licorice Jellies	9	334	–	83.3	HC
Nabisco Chuckles Marshmallow Eggs	10	392	0.6	95.4	HC
Nabisco Chuckles Nougat Centers	22	384	0.4	93.7	HC
Nabisco Chuckles Orange Slices	12	342	–	85	HC
Nabisco Chuckles Spearmint Leaves	13	337	–	84.1	HC
Nabisco Chuckles Spice Sticks and Drops	26	346	–	86	HC
Nabisco Chuckles Spice Strings	19	346	–	86.2	HC
Nabisco Cocoa-Covered Creme Wafer Sticks	11	547	4	64.3	HC
Nabisco Coco-Mello	5	462	3.8	69.8	HC
Nabisco Cocoanut Bars	11	482	5.7	67.6	HC
Nabisco Cocoanut Chocolate Chip	7	521	5.6	61.4	HC
Nabisco Cocoanut Family Favorite Cookies	30	492	6.9	68	HC
Nabisco Cocoanut Squares	7	423	4.0	80.6	HC
Nabisco Cookie Break Assorted Fudge	10	509	4.8	68.3	HC
Nabisco Cookie Break Fudge	10	513	4.6	68.6	HC
Nabisco Cookie Break Vanilla Creme	10	505	4.9	68.1	HC
Nabisco Corn Diggers Crackers	125	518	4.7	60.2	HC
Nabisco Cowboys and Indians Cookies	45	434	5.8	81.2	HC
Nabisco Crispy Clusters	6	402	2.7	86.5	HC
Nabisco Crown Peanut Bars	6	528	8.6	57.3	HC
Nabisco Crown Pilot Crackers	6	433	8.8	73.6	HC
Nabisco Dandy Cookies	9	426	4.3	71.6	HC
Nabisco Dandy Crackers	143	425	8.8	73.3	HC
Nabisco Dark Chocolate Cherries	6	398	1.4	77.3	HC
Nabisco Devil's Food Cakes	8	380	5.3	75.6	HC
Nabisco Doo Dads Crackers	200	480	9.3	61.4	HC
Nabisco Escort Crackers	24	483	6.5	64.2	HC
Nabisco Famous Chocolate Wafers	16	443	8.9	75.3	HC
Nabisco Fancy Crests	7	370	3.4	72.9	HC
Nabisco Fancy Grahams	7	509	5.4	66.8	HC
Nabisco Fancy Peanut Creme Cookies	9	541	9.5	58.1	HC

Food	Quantity	Calories	Protein Grams	Carbo-hydrates Grams	P/C/C Com-puter
Nabisco Fig Newtons	7	366	3.2	71.1	HC
Nabisco French Onion Crackers	42	481	7.4	67.3	HC
Nabisco Giant Sugar Daddy	2	398	2.2	87.7	HC
Nabisco Graham Crackers	14	431	7.1	76.8	HC
Nabisco Heydays Cookies	4	512	9.9	56.2	HC
Nabisco Holland Rusk Crackers	8	405	13.8	75.4	HC
Nabisco Home-Style Fudge	5	455	3.2	70.4	HC
Nabisco Iced Fruit Cookies	6	403	5.1	77	HC
Nabisco Ideal Peanut Bars	6	539	6.7	59	HC
Nabisco Jamaica Mints	17	397	–	97.1	HC
Nabisco Junior Mint Patties	40	416	1.8	81.1	HC
Nabisco Korkers Crackers	67	538	5.7	56.9	HC
Nabisco Lemon Snaps	26	437	5.3	81.1	HC
Nabisco Lemon Jumble Rings	7	464	5.2	74.9	HC
Nabisco Liberty Mints	17	398	–	97.1	HC
Nabisco Lorna Doone Cookies	13	485	6.2	67	HC
Nabisco Macaroons	7	501	4.4	66.8	HC
Nabisco Mallomars	8	459	4.3	66.6	HC
Nabisco Malted Milk Crunch	63	535	7.5	57.5	HC
Nabisco Marshmallow Cookies	13	411	5.2	73.4	HC
Nabisco Marshmallow Puffs	5	475	4.5	64.5	HC
Nabisco Marshmallow Twirls	3	440	3.8	72.3	HC
Nabisco Meal Mates, Onion Crackers	22	428	8.5	74.7	HC
Nabisco Meal Mates, Rye Crackers	22	421	7.8	75.7	HC
Nabisco Meal Mates, Sesame Crackers	22	447	10.5	68.3	HC
Nabisco Milk Chocolate Stars	36	536	5.4	55.6	HC
Nabisco Minarets	11	483	4.9	60	HC
Nabisco Mint Wafers	59	587	5.0	35.9	HC
Nabisco Mister Salty Dutch Pretzels	7	360	7.8	78.6	HC
Nabisco Mister Salty Pretzelettes	59	372	9.2	74.7	HC
Nabisco Mister Salty 3-Ring Pretzels	32	378	9.4	75.7	HC
Nabisco Mister Salty Veri-Thin Pretzels	19	387	9.3	78.1	HC
Nabisco Mister Salty Veri-Thin Sticks	333	371	9.4	78.3	HC
Nabisco Mystic Mint Cookies	6	521	5.6	63.3	HC
Nabisco National Arrowroot Crackers	21	466	6.2	73.0	HC
Nabisco Nilla Wafers	25	459	4.8	73.4	HC
Nabisco Nut Fudge	7	464	1.4	66.9	HC

Snack Foods

Food	Quantity	Calories	Protein Grams	Carbo-hydrates Grams	P/C/C Computer
Nabisco Nut Fudge Squares	7	464	1.4	66.9	HC
Nabisco Nutter Butter Cookies	7	488	8.6	64.8	HC
Nabisco Nutty Crunch	7	468	3.1	67.1	HC
Nabisco Oatmeal Cookies	6	466	5.8	70.5	HC
Nabisco Oatmeal Family Favorites Cookies	19	471	5.8	72.5	HC
Nabisco Old-Fashioned Ginger Snaps	14	420	5.9	76.8	HC
Nabisco Oreo Cookies	10	493	5	70.8	HC
Nabisco Oreo/Swiss Cookies	10	498	5.2	68.2	HC
Nabisco Oysterettes	125	425	8.8	73.3	HC
Nabisco Pantry Graham Cookies	8	506	5.1	69.5	HC
Nabisco Peanut Creme Cookies	15	520	10.7	59.2	HC
Nabisco Pecan Shortbread Cookies	7	541	4.2	61.2	HC
Nabisco Peppermint Patties	7	432	1.1	84.7	HC
Nabisco Pinwheels	3	458	2.8	69.1	HC
Nabisco Pom Poms Caramels	31	445	6.2	71.6	HC
Nabisco Premium Crackers, Unsalted	36	436	9.0	72.7	HC
Nabisco Premium Saltine Crackers	36	426	8.8	71.9	HC
Nabisco Pride Cookies	9	506	5.2	68	HC
Nabisco Raisin Fruit Cookies	7	381	4.7	81.1	HC
Nabisco Ritz Cheese Crackers	29	511	8.7	56.1	HC
Nabisco Ritz Crackers	30	494	6.7	62.9	HC
Nabisco Royal Clusters	6	484	9.2	46.2	HC
Nabisco Royal Lunch Crackers	9	470	7.2	67.5	HC
Nabisco Shapies Dip Delights Crackers	59	558	10.5	45.7	HC
Nabisco Shapies Shells Crackers	56	545	9.7	49.8	HC
Nabisco Sip 'n Chips Crackers	59	512	6.7	60.2	HC
Nabisco Sociables Crackers	48	474	10.2	62.9	HC
Nabisco Social Tea Biscuits	21	453	7.1	76.2	HC
Nabisco Social Tea Creme Biscuits	10	497	4.8	70.1	HC
Nabisco Spiced Wafers	10	434	4.9	77.6	HC
Nabisco Striped Shortbread Cookies	10	502	5	68.3	HC
Nabisco Sugar Babies	67	411	3.3	87.2	HC
Nabisco Sugar Daddy	3	400	2.1	87.1	HC
Nabisco Sugar Daddy Junior	8	398	2.2	87.7	HC
Nabisco Sugar Daddy Nuggets	8	398	2.2	87.7	HC
Nabisco Sugar Honey Maid Crackers	14	424	7.1	76.2	HC
Nabisco Sugar Mama	4	424	3.2	78	HC

Food	Quantity	Calories	Protein Grams	Carbo-hydrates Grams	P/C/C Computer
Nabisco Sugar Rings	7	468	5.7	72.5	HC
Nabisco Swiss Creme	10	507	5.8	64.2	HC
Nabisco Swiss 'n Ham Flings Crackers	56	553	9.0	45.4	HC
Nabisco Thin Mints	10	419	2.0	80.5	HC
Nabisco Triangle Thins Crackers	59	471	9.6	65.4	HC
Nabisco Triscuits	23	473	9.2	69.3	HC
Nabisco Twigs Crackers	36	498	10.5	57.7	HC
Nabisco Uneeda Crackers	19	431	9.1	73.3	HC
Nabisco Vanilla Snaps	35	437	5.8	80	HC
Nabisco Waverly Crackers	26	478	6.5	68.6	HC
Nabisco Welch's Chocolate Cherries	6	394	1.3	78.1	HC
Nabisco Welch's Cocoanut	3	437	3.3	72.1	HC
Nabisco Welch's Dark Chocolate Cherries	6	398	1.4	77.3	HC
Nabisco Welch's Frappe	3	437	2.4	77.6	HC
Nabisco Welch's Fudge	3	477	2.9	67.7	HC
Nabisco Wheat Thins	56	476	7.7	66.1	HC
Nabisco Whirligigs	16	416	2.3	82	HC
Nabisco Zuzu Ginger Snaps	26	426	4.6	80.5	HC
Nabisco Zwieback Crackers	14	428	12	74.4	HC
Pantry Pride Frosted Pastry	1	210	3	40	HC
Pantry Pride Unfrosted Pastry	1	200	2	35	HC
Peanut bars	3½ oz.	515	17.5	47.2	HC
Peanut brittle	3½ oz.	421	5.7	81.0	HC
Pepperidge Farm Bordeau Cookies	1	36	0.04	5.1	HC
Pepperidge Farm Brown Sugar Cookies	1	48	0.08	6.9	HC
Pepperidge Farm Brussels Cookie Components	1	42	0.5	4.6	HC
Pepperidge Farm Capri Cookies	1	82	0.09	9.7	HC
Pepperidge Farm Cardiff Cookie Components	1	18	0.2	18	LC
Pepperidge Farm Cheddar Cheese Goldfish Crackers	10	28	0.7	3.3	HC
Pepperidge Farm Cheese Toasted Thins Crackers	2	23	1	3.5	HC
Pepperidge Farm Chocolate Chip Cookies	1	52	0.06	6.2	HC
Pepperidge Farm Chocolate Nut Brownie	1	54	0.07	6.3	HC
Pepperidge Farm Chocolated Laced Pirouette Cookies	1	38	0.3	4.5	HC
Pepperidge Farm Cinnamon Sugar Cookies	1	52	0.06	7	HC

Food	Quantity	Calories	Protein Grams	Carbo-hydrates Grams	P/C/C Computer
Pepperidge Farm Dresden Cookies	1	83	0.08	10	HC
Pepperidge Farm Fudge Chip Cookies	1	51	0.06	6.7	HC
Pepperidge Farm Gingerman Cookies	1	33	0.04	5.4	HC
Pepperidge Farm Irish Oatmeal Cookies	1	50	0.06	7.1	HC
Pepperidge Farm Lemon Nut Crunch Cookies	1	57	0.07	6.4	HC
Pepperidge Farm Lemon Pirouette Cookies	1	37	0.3	4.4	HC
Pepperidge Farm Lido Cookies	1	91	1	10	HC
Pepperidge Farm Lisbon Cookie Components	1	28	0.3	3.3	LC
Pepperidge Farm Marquisette Cookie Components	1	45	0.5	5	HC
Pepperidge Farm Milano Cookies	1	62	0.07	7.2	HC
Pepperidge Farm Mint Milano Cookies	1	76	0.07	8.4	HC
Pepperidge Farm Molasses Crisp Cookies	1	30	0.04	4.3	HC
Pepperidge Farm Naples Cookies	1	33	0.04	3.7	HC
Pepperidge Farm Nassau Cookies	1	83	1.1	9.2	HC
Pepperidge Farm Oatmeal Raisin Cookies	1	55	0.8	7.5	HC
Pepperidge Farm Onion Goldfish Crackers	10	28	0.4	3.6	HC
Pepperidge Farm Onion Toasted Thins Crackers	2	23	0.6	4	LC
Pepperidge Farm Original Pirouette Cookies	1	37	0.3	4.4	HC
Pepperidge Farm Orleans Cookies	1	30	0.3	3.5	HC
Pepperidge Farm Parmesan Cheese Goldfish Crackers	10	28	0.8	3.4	HC
Pepperidge Farm Pizza Goldfish Crackers	10	29	0.5	3.6	HC
Pepperidge Farm Pretzel Goldfish Crackers	10	29	0.7	5	HC
Pepperidge Farm Rochelle Cookies	1	81	0.9	9.6	HC
Pepperidge Farm Rye Toasted Thins Crackers	2	21	0.7	3.9	LC
Pepperidge Farm Salted, Lightly, Goldfish Crackers	10	28	0.5	3.6	LC
Pepperidge Farm Sesame Garlic Goldfish Crackers	10	29	0.5	3.5	LC

Food	Quantity	Calories	Protein Grams	Carbo-hydrates Grams	P/C/C Computer
Pepperidge Farm Shortbread Cookies	1	72	0.8	8.3	HC
Pepperidge Farm Sugar Cookies	1	51	0.6	7	HC
Pepperidge Farm Tahiti Cookies	1	84	0.9	8.6	HC
Pepperidge Farm Venice Cookie Components	1	57	0.6	6.3	HC
Pepperidge Farm White Toasted Crackers	2	23	0.7	4	LC
Pillsbury Almond Danish Rolls w/icing	2	280	5	40	HC
Pillsbury Apple Cinnamon Coffee Cake Mix	1/8 cake	230	3	40	HC
Pillsbury Apple Cinnamon Cookie Mix	3	160	1	22	HC
Pillsbury Ballard Cinnamon Rolls w/icing	2	210	4	36	HC
Pillsbury Blueberry Coffee Cake Mix	1/8 cake	230	3	39	HC
Pillsbury Butter Pecan Coffee Cake Mix	1/8 cake	320	4	37	HC
Pillsbury Butterscotch Nut Cookie Mix	3	160	1	20	HC
Pillsbury Caramel Danish Rolls	2	310	4	41	HC
Pillsbury Caramel Sweet 'n Simple Rolls	1	250	3	38	HC
Pillsbury Cherry Coffee Cake Mix	1/8 cake	230	3	38	HC
Pillsbury Chewy Almond Cookie Mix	3	140	1	19	HC
Pillsbury Chocolate Chip Cookie Mix	3	170	1	23	HC
Pillsbury Cinnamon Country Coffee Cake	2 pieces	210	3	31	HC
Pillsbury Cinnamon Danish w/raisins	2	270	4	42	HC
Pillsbury Cinnamon Rolls w/icing	2	230	3	36	HC
Pillsbury Cinnamon Streusel Coffee Cake Mix	1/8 cake	250	3	41	HC
Pillsbury Cinnamon Sugar Cookie Mix	3	150	1	22	HC
Pillsbury Cinnamon Sweet 'n Simple Rolls	1	240	3	37	HC
Pillsbury Honey Sweet 'n Simple Rolls	1	250	3	39	HC
Pillsbury Hungry Jack Tastin' Cinnamon Rolls w/icing	2	290	4	39	HC

Food	Quantity	Calories	Protein Grams	Carbo-hydrates Grams	P/C/C Computer
Pillsbury Oatmeal and Chocolate Chip Cookie Mix	3	170	2	22	HC
Pillsbury Oatmeal Raisin Cookie Mix	3	180	2	27	HC
Pillsbury Orange Danish Rolls w/icing	2	260	4	42	HC
Pillsbury Orange Sweet 'n Simple Rolls	1	240	3	37	HC
Pillsbury Peanut Butter Cookie Mix	3	160	2	19	HC
Pillsbury Peanut Butter and Chocolate Chip Cookie Mix	3	150	2	17	HC
Pillsbury Sour Cream Coffee Cake Mix	1/8 cake	270	4	35	HC
Pillsbury Sugar Cookie Mix	3	160	1	21	HC
Pillsbury Swiss-Style Chocolate Chunk Cookies	3	160	1	21	HC
Popcorn:					
Oil-salt added	3½ oz.	456	9.8	59.1	HC
Plain	3½ oz.	386	12.7	76.7	HC
Sugar-coated	3½ oz.	383	6.1	85.4	HC
Pop-Tarts Frosted Blueberry Pastry	1	205	2.6	35.5	HC
Pop-Tarts Frosted Chocolate Fudge Pastry	1	215	3.8	35.2	HC
Pop-Tarts Frosted Dutch Apple Pastry	1	212	3.4	38.2	HC
Pop-Tarts Frosted Strawberry Pastry	1	201	2.7	37.4	HC
Potato chips	3½ oz.	568	5.3	50.0	HC
Potato sticks	3½ oz.	544	6.4	50.8	HC
Pretzels	3½ oz.	390	9.8	75.9	HC
Rye Krisp Original Crackers	3½ oz.	379	11.4	76	HC
Rye Krisp Seasoned Crackers	3½ oz.	412	10.8	68.5	HC
Sunshine Animal Biscuits	1	10	0.2	1.7	LC
Sunshine Applesauce Biscuits	1	83	0.9	11.9	HC
Sunshine Arrowroot Biscuits	1	16	0.2	3.0	LC
Sunshine Big Treat Biscuits	1	153	1.2	26.6	HC
Sunshine Butter-Flavored Biscuits	1	23	0.3	3.5	HC
Sunshine Cheez-its Biscuits	1	6	0.1	0.6	LC
Sunshine Cherry Cooler Biscuits	1	29	0.3	4.5	HC
Sunshine Chip-a-Roos Biscuits	1 large	63	0.6	7.7	HC
Sunshine Chocolate Chip Coconut Biscuits	1	80	0.8	9.7	HC
Sunshine Chocolate Fudge Biscuits	1	72	0.7	9.4	HC

Food	Quantity	Calories	Protein Grams	Carbo-hydrates Grams	P/C/C Com-puter
Sunshine Chocolate Snaps Biscuits	1	14	0.2	2.4	LC
Sunshine Cinnamon Toasts Biscuits	1	13	0.2	2.3	LC
Sunshine Coconut Bar Biscuits	1	47	0.5	6.2	HC
Sunshine Cream Lunch Biscuits	1	45	0.7	7.3	HC
Sunshine Cup Custard, Chocolate Biscuits	1	70	0.8	9.3	HC
Sunshine Cup Custard, Vanilla Biscuits	1	71	0.8	9.3	HC
Sunshine Dixie Vanilla Biscuits	1	60	0.9	13.1	HC
Sunshine Fig Bars	1	45	0.5	9.2	HC
Sunshine Ginger Snaps	1	24	0.3	4.4	HC
Sunshine Golden Fruit Biscuits	1	61	0.7	14.4	HC
Sunshine Grahams, Sweet Tooth Biscuits	1	45	0.4	6.4	HC
Sunshine HiHo Biscuits	1	18	0.2	2.1	LC
Sunshine Hydrox Cookies	1	48	0.4	7.1	HC
Sunshine Hydrox Mint Cookies	1	48	0.4	7.1	HC
Sunshine Hydrox Vanilla Cookies	1	50	0.4	7.1	HC
Sunshine Iced Animals Cookies	1	26	0.1	3.7	HC
Sunshine Iced Applesauce Cookies	1	86	0.9	11.9	HC
Sunshine Iced Aunt Sally Cookies	1	96	0.9	19.7	HC
Sunshine Iced Oatmeal Cookies	1	69	0.8	11.6	HC
Sunshine Krispy Biscuits	1	11	0.2	2	LC
Sunshine Lady Joan Iced Cookies	1	42	0.6	5.8	HC
Sunshine Lady Joan Plain Cookies	1	47	0.7	6.1	HC
Sunshine LaLanne Sesame Biscuits	1	15	0.3	1.8	LC
Sunshine LaLanne Soya Biscuits	1	16	0.2	1.9	LC
Sunshine Lemon Cookies	1	76	0.9	9.8	HC
Sunshine Lemon Coolers Cookies	1	29	0.3	4.5	HC
Sunshine Mallopuffs Cookies	1	63	0.5	12.2	HC
Sunshine Molasses and Spice Cookies	1	67	0.9	11.9	HC
Sunshine Oatmeal Cookies	1	58	0.7	8.9	HC
Sunshine Oatmeal Peanut Butter Cookies	1	79	1.2	10.5	HC
Sunshine Orbit Creme Sandwich Cookies	1	51	0.4	7	HC
Sunshine Oysters, Mini-Sized	1	3	0.1	0.6	LC
Sunshine Peanut Butter Patties	1	33	0.8	4.2	HC
Sunshine Scotties Cookies	1	39	0.6	5	HC

Snack Foods

Food	Quantity	Calories	Protein Grams	Carbo-hydrates Grams	P/C/C Computer
Sunshine Soda Crackers	1	20	0.4	3.3	LC
Sunshine Sprinkles Cookies	1	57	0.6	11.4	HC
Sunshine Sugar Cookies	1	86	1.0	11.9	HC
Sunshine Sugar Wafers Cookies	1	43	0.3	6.6	HC
Sunshine Sugar Wafers, Lemon Cookies	1	44	0.3	6.5	HC
Sunshine Toy Cookies	1	13	0.2	2.1	LC
Sunshine Vanilla Wafers, Small	1	15	0.2	2.2	LC
Sunshine Vienna Finger Sandwich Cookies	1	71	0.7	10.5	HC
Sunshine Yum Yums Cookies	1	83	0.5	10.4	HC
Toastettes Apple Pastry	3½ oz.	390	3.7	68.2	HC
Toastettes Blueberry Pastry	3½ oz.	388	3.7	69.8	HC
Toastettes Brown Sugar-Cinnamon Pastry	3½ oz.	399	4.3	67.5	HC
Toastettes Cherry Pastry	3½ oz.	385	3.9	69.2	HC
Toastettes Orange Marmalade Pastry	3½ oz.	382	4.4	67.8	HC
Toastettes Peach Pastry	3½ oz.	387	4.3	69.2	HC
Toastettes Strawberry Pastry	3½ oz.	388	3.7	69.1	HC
Wonder Bacon Rinds	3½ oz.	5	11.24	67.79	LC
Wonder Cheeze Twists	3½ oz.	539.91	8.29	49.42	HC
Wonder Corn Capers	3½ oz.	554.54	5.49	53.76	HC
Wonder Corn Chips	3½ oz.	556.65	5.28	51.62	HC
Wonder Cracker Jacks	3-3/8 oz.	510	9	90	HC
Wonder Onion Rings	3½ oz.	466.3	0.21	68.32	HC
Wonder Potato Chips	3½ oz.	552.3	5.14	48.61	HC
Wonder BBC Potato Chips	3½ oz.	538.16	5.25	48.51	HC
Wonder Taco Tort Chips	3½ oz.	504	6.05	55.75	HC
Wonder Tortilla Chips	3½ oz.	519.01	6.16	60.34	HC
Zwieback	3½ oz.	423	10.7	74.3	HC

Miscellaneous

Food	Quantity	Calories	Protein Grams	Carbo-hydrates Grams	P/C/C Computer
Aunt Jemima Regular Syrup	1 tbsp.	53	–	13.5	HC
Baking powder	3½ oz.	129	.1	31.2	HC
Carambola	3½ oz.	35	.7	8.0	HC
Chayote	3½ oz.	28	.6	7.1	LC
Cherimoya	3½ oz.	94	1.3	24.0	HC
Ginger root, crystallized (candied)	3½ oz.	340	.3	87.1	HC
Ginger root, fresh	3½ oz.	49	1.4	9.5	HC
Golden Griddle Regular Syrup	1 tbsp.	50	–	13	HC
Grapefruit peel, candied	3½ oz.	316	.4	80.6	HC
Honey	3½ oz.	304	.3	82.3	HC
Horseradish, raw	3½ oz.	87	3.2	19.7	HC
Karo Dark Corn Syrup	1 tbsp.	60	–	15	HC
Karo Imitation Maple	1 tbsp.	60	–	15	HC
Karo Light Corn Syrup	1 tbsp.	60	–	15	HC
Karo Pancake and Waffle Syrup	1 tbsp.	60	–	15	HC
Lemon peel, candied	3½ oz.	316	.4	80.6	HC
Lemon peel, raw	3½ oz.	–	1.5	16.0	LCR
Log Cabin Buttered Syrup	1 tbsp.	52	–	12.7	HC
Log Cabin Country Kitchen Syrup	1 tbsp.	51	–	13.1	HC
Log Cabin Maple-Honey Syrup	1 tbsp.	54	–	14	HC
Log Cabin Regular Syrup	1 tbsp.	46	–	12.2	HC
Longans, dried	3½ oz.	286	4.9	74.0	HC
Longans, raw	3½ oz.	61	1.0	15.8	HC
Lucerne Coffee Tone Creamer	1 tsp.	12	0	1	LC
Molasses, Barbados	3½ oz.	271	–	70	HC
Molasses, blackstrap	3½ oz.	213	–	55	HC
Old Manse Regular Syrup	1 tbsp.	53	–	13.2	HC
Orange peel, candied	3½ oz.	316	.4	80.6	HC
Orange peel, raw	3½ oz.	–	1.5	25.0	HCR
Peanut butter with small added fat, flavorings	3½ oz.	581	27.8	17.2	HC
Peanut spread	3½ oz.	601	20.3	22.0	HC
Poly Perx Non-Dairy Creamer	3½ oz.	140	1	11.94	HC
Salt, table	3½ oz.	0	0	0	–
Sandwich spread with chopped pickle	3½ oz.	379	.7	15.9	HC
Soyame Burger Aid	1 oz.	105	16	9	HC
Soyame Fortified	1 oz.	140	7	13	HC
Soyame Gran Burger	1 oz.	100	17	6	HC
Soyame Instant	1 oz.	130	7	12	HC
Soyame Instant Malt	1 oz.	130	7	12	HC
Soyame Regular	1 oz.	130	7	12	HC
Soyame Stripple Zips	1 oz.	130	14	2.9	HC

Food	Quantity	Calories	Protein Grams	Carbo-hydrates Grams	P/C/C Com-puter
Soyameat Beef-Like	1 oz.	30	2.5	1.7	HC
Soyameat Beef-Like, Diced	3½ oz.	150	19	3.2	HC
Soyameat Chicken-Like	1 oz.	35	2.8	1	HC
Soyameat Diced Chicken-Like	1 oz.	35	2.8	1	HC
Soyameat Fried Chicken-Like	1 piece	65	4.5	0.5	HC
Soyameat Salisbury Steak-Like	1 piece	160	12	7	HC
Sugars:					
Anhydrous	3½ oz.	366	0	99.5	HC
Brown	3½ oz.	373	0	96.4	HC
Crystallized	3½ oz.	335	0	91	HC
Granulated	3½ oz.	385	0	99.5	HC
Maple	3½ oz.	348	0	90	HC
Powdered	3½ oz.	385	0	99.5	HC
Syrups:					
Cane	3½ oz.	263	0	68	HC
Cane and maple	3½ oz.	252	0	65	HC
Light corn table blend	3½ oz.	290	0	75	HC
Maple	3½ oz.	252	—	65	HC
Sorghum	3½ oz.	257	—	68	HC
Tillie Lewis Tasti-Diet Sweetnin'	1 tsp.	0	0	0	—
Tomato paste, canned	3½ oz.	82	3.4	18.6	HC
Tomato puree, canned	3½ oz.	39	1.7	8.9	HC
Vinegar, cider	3½ oz.	14	Trace	5.9	LC
Vinegar, distilled	3½ oz.	12	—	5	LC
Worthington Beef-Like	1 slice	55	6	1.4	LC
Worthington Chicken-Like	1 oz.	75	6	0.8	HC
Worthington Chic-ketts	1 oz.	55	7	1.1	HC
Worthington Chili	¼ can	165	14	14	HC
Worthington Choplet Burger	1/3 cup	120	13	11	HC
Worthington Choplets	1	50	8	2	LC
Worthington Corned Beef-Like	1 slice	35	2.5	1.2	LC
Worthington Croquettes	1	70	5	5	LC
Worthington Fillets	1	115	10	3.5	HC
Worthington FriPats	1	200	14	11	HC
Worthington Fry Sticks	1	120	14	5	HC
Worthington Non-Meat Balls	1	35	3.2	1.1	LC
Worthington Numere	½-inch slice	160	9	7	HC
Worthington Prosage	3-1/8-inch slice	80	6	3.5	HC
Worthington Protose	½-inch slice	175	19	6	HC
Worthington Salisbury Steak-Like	1 slice	110	9	6	HC
Worthington Sandwich Spread	3 tbsp.	80	5	3.5	HC
Worthington Saucettes	1 link	40	3.3	0.5	LC
Worthington Smoked Beef-Like	1 slice	15	1.6	0.4	LC

Food	Quantity	Calories	Protein Grams	Carbo-hydrates Grams	P/C/C Com-puter
Worthington Smoked Turkey-Like	1 slice	45	3.6	1.2	HC
Worthington Stripples	1	15	1.6	0.3	LC
Worthington 209	1 oz.	35	2.5	1.1	LC
Worthington Vegetable Scallops	1	30	5	0.9	LC
Worthington Vegetable Steaks	1 piece	30	5.8	0.4	LC
Worthington Vegetarian Burger	1/3 cup	100	12	3.4	HC
Worthington Veja-Links	1 link	75	4.1	1.2	HC
Worthington Wham	1 oz.	45	4.2	1.3	HC
Worthington Wham	1 slice	55	6	1.2	HC

Baby Foods

Food	Quantity	Calories	Protein Grams	Carbo-hydrates Grams	P/C/C Computer
Applesauce	3½ oz.	72	.2	18.6	HC
Applesauce-apricots	3½ oz.	86	.3	22.6	HC
Bananas (with tapioca or cornstarch) strained	3½ oz.	84	.4	21.6	HC
Bananas-pineapple (with tapioca or cornstarch)	3½ oz.	80	.4	20.7	HC
Barley, added nutrients	3½ oz.	348	13.4	73.6	HC
Beans, green, canned	3½ oz.	22	1.4	5.1	LC
Beef heart	3½ oz.	93	13.5	.4	HC
Beef, junior	3½ oz.	118	19.3	0	HC
Beef noodle dinner	3½ oz.	48	2.8	6.8	HC
Beef, strained	3½ oz.	99	14.7	0	HC
Beef-vegetables	3½ oz.	87	7.4	6.0	HC
Beets, strained, canned	3½ oz.	37	1.4	8.3	LC
Carrots, canned	3½ oz.	29	.7	6.8	LC
Cereal, egg yolk, and bacon	3½ oz.	82	2.9	6.6	HC
Chicken	3½ oz.	127	13.7	0	HC
Chicken noodle dinner	3½ oz.	49	2.1	7.2	HC
Chicken-vegetables	3½ oz.	100	7.4	7.2	HC
Custard pudding, all flavors	3½ oz.	100	2.3	18.6	HC
Egg yolks, strained	3½ oz.	210	10.0	.2	HC
Egg yolks with ham or bacon	3½ oz.	208	10.0	.3	HC
Fruit dessert with tapioca	3½ oz.	84	.3	21.5	HC
Fruit pudding	3½ oz.	96	1.2	21.6	HC
High protein, added nutrients	3½ oz.	357	35.2	48.1	HC
Lamb, junior	3½ oz.	121	17.5	0	HC
Lamb, strained	3½ oz.	107	14.6	0	HC
Liver, strained	3½ oz.	97	14.1	1.5	HC
Liver-bacon, strained	3½ oz.	123	13.7	1.3	HC
Macaroni-tomato-meat	3½ oz.	67	2.6	9.6	HC
Mixed, added nutrients	3½ oz.	368	15.2	70.6	HC
Mixed vegetables (including soup)	3½ oz.	37	1.6	8.5	LC
Oatmeal, added nutrients	3½ oz.	375	16.5	66.0	HC
Peaches	3½ oz.	81	.6	20.7	HC
Pears	3½ oz.	66	.3	17.1	HC
Pears-pineapple	3½ oz.	69	.4	17.6	HC
Peas, strained	3½ oz.	54	4.2	9.3	HC
Plums-tapioca, strained	3½ oz.	94	.4	24.3	HC
Pork, strained	3½ oz.	118	15.4	0	HC
Pork, junior	3½ oz.	134	18.6	0	HC
Prunes-tapioca	3½ oz.	86	.3	22.4	HC
Rice, added nutrients	3½ oz.	371	6.6	80.0	HC
Spinach, creamed	3½ oz.	43	2.3	7.5	HC
Split peas-vegetables and ham or bacon	3½ oz.	80	4.0	11.2	HC

Food	Quantity	Calories	Protein Grams	Carbo-hydrates Grams	P/C/C Computer
Squash	3½ oz.	25	.7	6.2	LC
Sweet potatoes	3½ oz.	67	1.0	15.5	HC
Teething biscuit	3½ oz.	378	11.1	78.0	HC
Tomato soup, strained	3½ oz.	54	1.9	13.5	HC
Turkey-vegetables	3½ oz.	86	6.7	7.6	HC
Veal, junior	3½ oz.	107	18.8	0	HC
Veal, strained	3½ oz.	91	15.5	0	HC
Veal-vegetables	3½ oz.	63	7.1	5.1	HC
Vegetables-bacon-cereal	3½ oz.	68	1.7	8.7	HC
Vegetables-beef-cereal	3½ oz.	56	2.7	7.6	HC
Vegetables-chicken-cereal	3½ oz.	52	2.1	7.7	HC
Vegetables-ham-cereal	3½ oz.	64	2.8	8.3	HC
Vegetables-lamb-cereal	3½ oz.	58	2.2	7.7	HC
Vegetables-liver-bacon	3½ oz.	57	2.4	7.5	HC
Vegetables-liver-cereal	3½ oz.	47	3.1	7.8	HC
Vegetables-turkey-cereal	3½ oz.	44	2.1	7.2	HC